Earth Matters

The Earth Sciences, Philosophy, and the Claims of Community

Robert Frodeman, Editor
University of Tennessee, Chattanooga

Contributing Authors

Victor R. Baker
University of Arizona

Karim Benammar
Kobe University, Japan

Albert Borgmann
University of Montana

Bruce V. Foltz
Eckerd College

Alphonso Lingis
Penn State University

W. Scott McLean
University of California, Davis

Carl Mitcham
Colorado School of Mines

Eldridge M. Moores
University of California, Davis

Max Oelschlaeger
Northern Arizona University

Brian Polkinghorn
Nova Southeastern University

David A. Robertson
University of California, Davis

Daniel Sarewitz
Columbia University

Kristin Shrader-Frechette
University of Notre Dame

Christine Turner
U.S. Geological Survey

Peter Warshall
Whole Earth Magazine

Richard S. Williams, Jr.
U.S. Geological Survey

Prentice Hall
Upper Saddle River, NJ 07458

Library of Congress Cataloging-in-Publication Data

Earth matters: the earth sciences, philosophy, and the claims of community / Robert
Frodeman, editor; contributing authors, Victor R. Baker ... [et. al.].
    p. cm.
  Includes bibliographical references and index.
  ISBN 0-13-011996-2
    1. Earth sciences--Philosophy. 2. Environmental sciences.. I. Frodeman, Robert. II.
  Baker, Victor R.

  QE33 .E34 2000
  550'.12--dc21

                                                                    99-057001

Executive Editor: Daniel Kaveney
Assistant Editor: Amanda Griffith
Production Editor/Page Layout: Kim Dellas
Manufacturing Manager: Trudy Pisciotti
Buyer: Michael Bell
Art Director: Jayne Conte
Cover Designer: Bruce Kenselaar
Cover Photo Credit: Fred Hanselmann-Rocky Mountain Photography
Editorial Assistant: Margaret Ziegler

Printed in the United States of America

10 9 8 7 6 5 4 3 2 1

ISBN 0-13-011996-2

Prentice-Hall International (UK) Limited, *London*
Prentice-Hall of Australia Pty. Limited, *Sydney*
Prentice-Hall Canada Inc., *Toronto*
Prentice-Hall Hispanoamericana, S.A., *Mexico City*
Prentice-Hall of India Private Limited, *New Delhi*
Prentice-Hall of Japan Inc., *Tokyo*
Pearson Education Asia Pte. Ltd.
Editora Prentice-Hall do Brasil, Ltda., *Rio de Janeiro*

To Christine
and Carl

Contents

Preface

SHIFTING PLATES: THE NEW EARTH SCIENCES

1.

Changes are afoot within the discipline of geology. One sign of this shift is the controversy over what the field should be called. Do we stay with the traditional term, *geology*? Or embrace a new name, such as *geoscience*, *Earth science*, or *Earth systems science*? I prefer the first term: despite its implication of referring exclusively to the solid Earth, geology contains hidden resonances. In ancient Greek, *Gé* or *Gaia* meant *earth* in the sense of the soil, but also Mother Earth, the sheltering source of life, as well as one's country or homeland. Geology may then be taken as the search for a *logos* of *Gaia*—an account of Mother Earth, the planet that we call home.

The point I wish to make, however, is not one of nomenclature, but to alert readers to what is at stake in this debate over names. For the Earth sciences (to use the most common term) today find themselves at the center of a number of societal changes—concerning the relation between science and community, the limits of science for solving the problems we face, and the creation of new, interdisciplinary models for problem solving.

Earth Matters provides a forum for addressing these questions. The essays of this volume—all published for the first time—represent a wide spectrum of opinion on what the Earth sciences are and should become, and more generally, what our relation to the Earth is and should be. Written by scientists, humanists, and practitioners of public policy, *Earth Matters* provides support for a wide range of views concerning the role of the Earth sciences in society.

The questions raised by these essays include: How has the role of the Earth scientist—and the scientist in general—changed in recent years? What new demands or obligations has society placed on science? Can science and technology resolve the environmental difficulties we face, or must we recognize the limits of scientific and technical solutions to our problems, and develop new approaches that combine the insights of the social sciences and the humanities with those of the sciences? Is our relation to the Earth exclusively defined in terms of utilitarian concerns, or is our response to the Earth as much a matter of aesthetics and theology as it is of economics? If the latter is the case, what consequences will this have for the future of the Earth sciences? Finally, how are we to understand the relation between scientific research and the needs of communities, whether on the local, regional, national, or international scale?

2.

The idea of promoting this conversation about the Earth sciences and the needs of community has a particular background that it is perhaps useful to mention. First educated in history and philosophy, after a few years of teaching philosophy at the college level I became dismayed at the sterility of much academic philosophical discussion. Motivated by the desire to bring philosophy down to Earth, I returned to graduate school to pursue a master's degree in geology.

In the course of this return to the status of a student, I became aware of serious philosophical discussions among Earth scientists themselves. These discussions fell into three broad categories. First, Earth scientists voiced concerns with the nature of the scientific method as it applied to their work. Although many geologists were successfully reducing geology to geophysics and geochemistry, others found that their work did not easily fit within the standard model for the scientific method. These geologists played a role closer to that of Sherlock Holmes, dependent on the interpretation of clues, rather than being able to test their hypotheses in the laboratory. The historical nature of geology, as well as the inability to duplicate in the lab the spatial or temporal scales of the field, meant that geologists relied on skills that were often closer to those of the detective, the physician, or the literary critic, than to the chemist.

Second, many of the Earth scientists I spoke to were concerned with the relation of science to community concerns. Some were confident that scientific information and perspectives, once grasped by the public, would necessarily lead to sound public policy. Others, however, worried that this assumption misconceived the nature of both the scientific and the political realms. On this reading, emphasizing the political relevance of scientific research was liable to raise false hopes among the public. The danger here was that science would be asked to answer what were essentially political questions, that is, questions about societal values that must be adjudicated through the political process.

Third, I was struck by what can be called the philosophical nature of much of scientific research. Many of the scientists I came in contact with were motivated by the pure love of learning—the sense of wonder and the experience of beauty that comes from understanding the nature of things. In this way, many Earth scientists seemed closer to the traditional ideal of the natural philosopher than to the caricature of the passionless, objective scientist that I grew up with. This insight turned upside-down the meaning of terms like *practical*, for I realized that the philosophical and political aspects of the Earth sciences are as important to the public as are the economic benefits of the Earth sciences. The discovery of a new species of dinosaur has no economic implications, but our fascination with such marvels points to the importance of the philosophical aspects of science.

The more I came to know professional geologists, the more I recognized that we shared common philosophical concerns and interests. I began to attend and present papers at geological conferences such as the annual meetings of the Geological Society of America, and I was hired as a consultant to the United States Geological Survey on matters of sciences, ethics, and public policy. The present collection has grown out of this dialogue between the Earth sciences and philosophy.

3.

Long ignored by the humanities, and traditionally seen by society as simply the supplier of raw materials for the industrial machine, the Earth sciences today are moving to the center of public consciousness and conversation. The reasons for this are clear.

Resource scarcity, global climate change, the loss of biodiversity, pollution, and the expansion of a Western consumerist lifestyle around the globe means that the limits of resources and ecosystems have become crucial to societal debates. In brief, after a long period of ignoring it, we have discovered that the Earth matters.

To this point, geology has played a marginal role within the humanities. With very few exceptions, the philosophy of science has ignored the discipline. One finds no philosophy of geology as one does a philosophy of physics and of biology. The two main schools of contemporary philosophy, Analytic and Continental, have ignored geology. Both have assumed (few thought to argue the point) that an examination of geology was unnecessary for understanding the nature of science.[1]

Philosophy's neglect of geology is exemplified by the lack of attention given to the concept of geologic time. The discovery of "deep" or geologic time parallels in importance the widely acknowledged Copernican Revolution in our conception of space. And in fact, the awareness of history and temporality play a quite prominent role within contemporary Continental philosophy. Nevertheless, philosophers have ignored the decisive role played by the Huttonian Revolution in reshaping our sense of time and history. In general, geology has been viewed as a derivative science with no distinctive method of its own, reducible to the principles of chemistry and physics. Geology suffered from a host of problems that undercut its claims to knowledge: the incompleteness of geologic data; the lack of experimental control that is possible in the laboratory-based sciences; and the great spans of time required for geologic processes to occur, making direct observation difficult or impossible.

These factors left the field a less-than-ideal candidate for philosophic consideration. In fact, the philosophy of science has traditionally viewed physics as the paradigmatic science. Physics was the first science to establish itself on a firm footing, exemplifying the true nature of science as certain, precise, and predictive knowledge of the world. Since the seventeenth century, all other sciences (and the humanities as well) have been judged in terms of how well they live up to these standards.

Yet, there is another way of looking at the Earth sciences: seeing them as offering a model of human rationality more in keeping with the realities we face in our personal and public lives. The Earth sciences do have a distinctive method of reasoning: one that is deliberative rather than simply calculative, interpretive rather than purely factual, and historical rather than experimental—again, like our own personal and public lives. The reasoning process typical of the Earth sciences thus offers us a middle way between the often unrealistic standards of the lab sciences—based as they are on the essentially falsifying nature of the controlled experiment—without slipping into the no-nothing-ness, fundamentalism, and blind deference to authority that is the antithesis of rational-

---

[1]Typical conclusions by philosophers concerning the status of geology are those of Nelson Goodman: "In conclusion, then, the Principle of Uniformity dissolves into the principle of simplicity that is not peculiar to geology but pervades all science and even daily life." "Uniformity and Simplicity," in C.C. Albritton, et al., eds., *Uniformity and Simplicity: A Symposium on the Principle of Uniformity in Nature*: New York: Geological Society of America Special Paper 89 (1967), p. 99; and Richard A. Watson, "Geology is a science just like other sciences, for example physics or chemistry." In "Explanation and Prediction in Geology": *Journal of Geology*, v. 77, p. 488.

There have been some exceptions to this attitude. N. Oreskes, K. S. Shrader-Frechette, and K. Belitz have addressed the question of the limits of modeling in a 1994 essay, "Verification, Validation, and Confirmation of Numerical Models in the Earth Sciences," *Science* 263, pp. 641–646. N., (1994). Shrader-Frechette also offers an analysis of the assumptions underlying the storage of nuclear waste in *Burying Uncertainty: Risk and the Case Against Geological Disposal of Nuclear Waste* (Berkeley: University of California Press, 1993). John Sallis's *Stone* (Bloomington: Indiana University Press, 1994) is a recent exception to the general neglect of our lithic substrate within Continental philosophy.

ity. Geology is a preeminent example of a synthetic science, combining a variety of logical techniques in the solution of its problems. The geologist exemplifies what the French anthropologist Lévi-Strauss called the *bricoleur*, the thinker whose intellectual toolbox contains a variety of tools that he or she selects as is appropriate to the job at hand. The Earth sciences, then, can be viewed as a bridge discipline between the laboratory sciences and the modes of reasoning characteristic of the humanities.

4.

Geology today is caught in a kind of cultural tectonics, as the relations between the Earth sciences and the rest of society are being transformed. Geoscientists are being asked to take on new roles and responsibilities that often extend beyond their specific disciplinary expertise. In fact, many of these roles involve issues that are fundamentally political or philosophical in nature.

Take, for instance, the case of the disposal of the nation's nuclear waste. At Yucca Mountain, Nevada, Earth scientists are being asked to certify the safe storage of tens of thousands of tons of highly radioactive material for 10,000 years—a length of time greater than the recorded history of human culture. With such questions, the lines dividing science from epistemology, ethics, politics, and metaphysics blur. How confident can anyone be about what might occur over the next 10,000 years? And, what is the nature of our responsibility to generations unborn? Other Earth science issues—global climate change, the loss of biodiversity, resource shortages, pollution, and natural hazards—are no less challenging to our traditional understanding of the role of the sciences in society.

But rather than being an exception, the situation of the Earth sciences today is exemplary for the future of the sciences, and I believe for the humanities as well. For the Earth sciences simply make clear what is becoming apparent everywhere: The nature of knowledge is changing, as is the place of knowledge in society. In today's culture, information *disseminates*: knowledge can no longer be treated as existing in discrete packets. Instead, for both logical and political reasons, every discipline must show how its insights fit with the concerns of society. The Earth sciences, then, provide us with an image of the challenges that all disciplines will face in the twenty-first century.

In response to these new conditions, there are signs that a consensus is emerging within the geoscience community. This consensus recognizes that the challenge of global change calls not only for advances in scientific research and methodology, but also for an enlarged sense of stewardship, ethics, and cross-disciplinary integration among the disciplines. In other words, the geoscience community has found that it must now rethink fundamental questions concerning its role within, and responsibilities to, society.

It is certainly true that changes in technology—for example, faster computers, and the rise of integrated observation systems—have altered the possibilities for research within the Earth sciences and for all other disciplines. But the fundamental fact ushering in the new era is this: For the first time in the history of the planet, humans are a major geologic force. We now affect climate and biodiversity in unprecedented ways at the local, regional, national, and global levels. Furthermore, these changes result from our use of energy and mineral resources at levels that are unlikely to be sustainable over the long term. Accordingly, the geosciences will face greater challenges and will have a larger role to play than ever before, in both the future of human well-being and in the health of the planet.

5.

The transformation of the discipline of geology is being driven by at least two forces. First, the reformation of geology became inevitable once we entered an age of ecological and geological scarcity. The concept of scarcity invoked here must not be thought of pointing toward a pure fact of nature. Environmentalists have too often looked to the Earth sciences for the identification of positive and inescapable limits that will force Western societies to radically alter their lifestyle. Such a concept of scarcity or limit is easily dismantled—time and again proven wrong by history, as expected shortages of energy, metals, or food are overcome. Humans are simply too resourceful, too capable of modifying their behavior or inventing new technologies, to easily fall into this trap.

Geologic scarcity *is* real. But the geologic scarcity that we will experience will be as much a cultural as a natural phenomenon. Geologic scarcity will be defined by the interplay of physical limits (always uncertain, and subject to change through new discoveries and technological advance) and a complex range of cultural limits, involving factors such as technology, economics, ethics (questions of justice), aesthetics (quality of life issues), and theology (a sense of the sacred). Consider the example of petroleum. Rather than simply running out of oil, we will eventually change our patterns of energy consumption because of our unwillingness to accept the consequences of its continued use: traffic jams, polluted air, communities given over to cars, compromised relations to foreign nations, and damage to beautiful and fragile places (e.g., the Arctic, or the California coast). These limitations will then prompt both technological advances that will allow us to meet our energy needs in new ways, and lifestyle changes that allow us to live more lightly on the Earth.

The second reason for suggesting that we are witnessing the birth of a new kind of geology turns on what could be called—if a touch of hyperbole is allowed—the death of the natural. Technology today has become transformative. Reaching deep within the structure of reality, we create fibers and materials that are truly manmade, colors never before seen by the eye, and forms of life that have never before existed. We are on the verge of manipulating the genetic stock of our species. Through all these efforts we are wiping away the very distinction between the artificial and the natural. The fossil in the rock shop—is it real, or a replica? The photograph of the sunset— were its colors changed on the computer, or possibly in the processing? Even ecosystems can now be restored so that the visitor (and, in some cases, the expert) cannot tell that the area was ever disturbed.

It must be emphasized that our hesitancy to this manipulation of reality is more than just practical in nature. The possibility that a bacterium engineered for one purpose can come to play another, less benign role in our lives is real enough. But there is also a growing sense that the natural has a status and a claim upon our attention all its own. We want natural or near-natural landscapes, areas wholly or largely untouched by human manipulation, because such landscapes preserve a distinction that we rely upon—a sense of limit, or of the sacred.

Thus, in all the excitement and celebration accompanying the tremendous expansion of human prowess through technology, I detect a note of dismay. This dismay is directed at the loss of the natural—a way that things not only are, but in some sense should be. If I am correct, this urge for things beyond human manipulation finds its greatest expression in the contemporary apotheosis of nature and wilderness. Earth science facts and perspectives, then, also function as a type of geotheology, as we search the rocks and fossils for something that transcends our overly built world.

But these are merely two of the most salient factors that suggest that the Earth sciences are destined to become much more culturally prominent in the twenty-first century. If this is correct, this represents a tremendous opportunity to the Earth science community—if it is able to adapt to the changing nature of the demands society will place upon it. In the twenty-first century, the Earth sciences need to become a discipline that says no as well as yes to society: not only enabling our plans for industrial and technological development, but also describing the limits to our activities, as they manifest themselves through geologic hazards, resource scarcity, and ecosystem stress.

Such as role is, of course, very different from the traditional, nineteenth- and twentieth-century concept of the scientist to which we have grown accustomed. On this view, the scientist can and must remain objective, completely divorced from cultural or political commitments. The scientist's work was finished when the science was competently completed by standards internal to the scientific process. Under conditions of geologic scarcity, however, the Earth scientist must become a *political* or *public* scientist, guided by community needs at the same time that he or she provides counsel and advice.

Conversely, the recognition that geology has a social, political, and philosophical role to play also opens new prospects for disciplines such as history, philosophy, and literary studies. Students and scholars in the humanities will find that they too have a place at the table in contemporary societal debates—*if* they learn to bring their work down to Earth, making it accessible to nonspecialists. Such a marriage of scientific, political, and philosophical perspectives within the Earth sciences holds the promise of healing the split between the two cultures of science and the humanities.

6.

One of the inherent features of philosophical reflection is that it is always crossing boundaries. Therefore, while the essays collected here are organized around the three themes of the logic of the Earth sciences, the Earth sciences and society, and the philosophical implications of the Earth sciences, each of the essays address all of these themes to one degree or another. All of these essays honor science, while calling for a wider knowledge or wisdom; an understanding of human abilities and limits, and a consideration of the nature of goodness, truth, and beauty. Such reflection is an ongoing process that science both builds on and can awaken—but which ultimately goes beyond science.

This collective conversation about Earth matters aims to promote this discussion as a scholarly end in itself. But each of us, no matter what our disciplinary specialty, is in the first instance a citizen and reflective human being. In a world in which the information and perspectives of the Earth sciences play an increasingly important role, both the academic community and the general public will be well served by a dialogue on the changing role of the Earth sciences in society.

More particularly, this collection may serve as a textbook or supplementary text in various courses in the sciences and the humanities, at both the undergraduate and graduate levels. Students in a general education course in the Earth or environmental sciences want to explore the meaning of what they are learning. And, professors are increasingly appreciative of the need to deepen public understanding of their disciplines by including material that reflects on the societal importance of the sciences. In the fields of environmental studies and environmental philosophy, this collection may also serve as a primary or supplementary text. Such courses must include reflection not just on the environment, but also on the sciences that mediate that environment to us.

My thanks to the following individuals, without whom this project would not have come to fruition: Daniel Kaveney, Geoscience Editor at Prentice Hall; Carl Mitcham, Department of Liberal Arts and International Studies, Colorado School of Mines; Christine Turner, US Geological Survey; Ian MacGregor and Michael Mayhew, National Science Foundation; Erle Kauffman, Indiana University; Hartmut Spetzler, University of Colorado; and Kathryn Mutz and the Natural Resources Law Center at the University of Colorado, whose El Paso Fellowship, funded by the El Paso Energy Foundation, provided me with the release time needed to complete this project.

Robert Frodeman
Grand Canyon Semester
Flagstaff, AZ

ROCK LOGIC: THE NATURE OF THE EARTH SCIENCES

1

Conversing With the Earth:
The Geological Approach
to Understanding

Victor R. Baker

Victor R. Baker is Regents Professor and Head of the Department of Hydrology and Water Resources at the University of Arizona. He was also the 1998 President of the Geological Society of America. Baker has authored over 200 scientific research papers and several books concerned with paleohydrology, geomorphology, planetary geology, the history and philosophy of Earth science, and the role of the Earth sciences in public policy.

In this essay, Baker offers an account of two approaches to the scientific examination of nature, what he calls "Earth-directed" and "theory-directed" science. In ways similar to the work of the twentieth-century German philosopher Martin Heidegger, Baker argues that the process of controlled experimentation in science is itself theory-laden, in that it assumes or imposes lawfulness upon the world. In contrast, "Earth-directed" science is a "conversation with the Earth" where the geologist attends to the signs or messages provided by the natural phenomena themselves. This approach, while also theory-laden, is the more fundamental moment within scientific research, for the data used in the testing of a scientific model or theory are themselves the results of semiotic acts where natural phenomena are interpreted. While making clear that mathematical modeling has a central role to play in understanding environmental change, Baker emphasizes that these models depend upon the reading of theory-laden signs (what he calls "indexicals"), rather than proxies that are exact and unequivocal reflections of the world.

How can a serious scientist claim that one converses with the Earth? Surely this is but a poetic metaphor. Obviously, a scientist can speak to the world, but there is no way for the world to speak back. This point was made abundantly clear by the top philosopher on *Time* magazine's recent list of the leading hundred minds of the twentieth century: Ludwig Wittgenstein. Wittgenstein begins his book *Tractatus Logico-Philosophicus* with this presumption: "The world is all that is the case." The world consists only of facts and is solely determined by facts. Wittgenstein's brilliant logical analysis, written early in his career, holds that it is language that comes to reflect the structure of the world.

The idea that our scientific language, or ideas, or models come to mirror nature is one of those fundamental assumptions that most scientists presume without thinking about the matter. It is precisely in such concerns that philosophy of science finds a role

2

amid the hurried pace of scientific activity. The emphasis upon the logical structure of our language for expressing statements about the world is considered necessary for the ultimate precision of checking those statements against the measured facts of the world.

Modern physics has discovered a difficulty of making definitive measurements at the level of the quantum world. It has become clear that the old idea of models mirroring the world has fundamental limitations. The point is made by a question once put to the physicist Niels Bohr. Asked whether the elegant theory of quantum mechanics mirrored the world, Bohr replied: "There is no quantum world. There is only an abstract quantum description. It is wrong to think that the task of physics is to find out how nature is. Physics concerns what we can say about nature."[1]

The problems of representation and language mediating between mind and nature constitute a great unresolved philosophical issue. Perhaps the most important example of this issue in geology is provided by doctrines introduced by Charles Lyell, and commonly accorded the label "uniformitarianism." It was Lyell's agenda to promote uniformitarianism for producing logical validity in what could be said about the Earth. Lyell's logical presumptions were questioned at the time of their introduction in the 1830s by Cambridge mineralogist William Whewell.[2] Whewell criticized Lyell's uniformitarian doctrine for its presumptions of both (1) the nature of causes operative in nature, and (2) the nature of proper scientific reasoning as to those causes. In his 1837 book *History of the Inductive Sciences*, Whewell said of Lyell's method that it "is not, we may suggest, the temper in which science ought to be pursued." Instead of the scientist saying what causes are appropriate to consider, as advocated by Lyell, Whewell argued that "The effects must themselves teach us the nature and intensity of the causes which have operated."

How can effects teach us their causes? Again, we seem to be asking the Earth to speak to us. Perhaps we need to contrast, for discussion purposes, two styles of science, one in which the Earth speaks to us, call it "Earth-directed" science, and another in which we, more specifically our language or theories, speak to the Earth, call it "theory-directed" science. Obviously, this contrast is a bit extreme and artificial, but it may illustrate some ideas that help with the notion of "conversing with the Earth."

Theory-directed science is best exemplified in the experimental/theoretical methodology of classical physics. Indeed, much of the philosophy of science is written as though the words "physics" and "science" are interchangeable. The approach of physics is conceptual, seeking universal classes of phenomena that can be generalized by means of the underlying physical laws that are presumed to govern nature. The abstract laws, theories, and relationships of physics must be verified against measured reality through controlled experimentation. The facts of nature are expressed through numerical measures comparable to those generated by the theoretical representation of underlying laws. Comparison of quantitative measurements to the theoretical predictions results in scientific validation of theories.

Table 1.1 compares theory-directed science to Earth-directed science. The latter does not focus on idealized theories verified in experimental laboratories. Instead, the prime concern of Earth-directed science is with realized phenomena observed in the natural world, uncontrolled by artificial constraints. By not limiting oneself to the world amenable to mathematical analysis, the naturalistic scientist takes the world as it is. Rather than general principles of universal application, concrete particulars are the

Table 1.1 **COMPARISON OF REASONING STYLES IN EARTH SCIENCE**

	Theory-Directed	**Earth-Directed**
Basis	Define elements of nature (systems) capable of controlled study	Take the world (nature) "as it is"
Goal	Develop theories that explain Earth	Develop understanding of Earth
Emphasis	Idealizations: general principles presumed to apply at all times to all places	Real phenomena: concrete particular happenings, past and present
Characteristics	Experimental, predictive, mathematical	Experimental, historical, observational
Methods	Controlled experimentation and model simulation	Observation (in the field) to stimulate hypotheses
Tools of Study	Facts and theories	Signs
Role of "Data"	Verification (validation) of model predictions	Signs providing indices of causal processes
Types of Inference	Deductive analysis (rigorous and elegant) and inductive synthesis (for theory confirmation)	Retroductive (abductive) synthesis followed by deduction and induction
Role of Logic	Valid reasoning in regard to what we can say about Earth	Fruitful reasoning emphasizing what Earth says to us

focus of attention. The richest source of such reality is the various evidence of happenings in the past. Observations of past phenomena are revealed as signs, providing a language for what Hans Cloos termed a "conversation with the Earth."[3] The observations do not serve primarily the menial function of model validation. Rather, as signs, they lead the investigator to create hypotheses. Hypothetical reasoning, therefore, provides a kind of logic, or inference, through which one can distinguish the naturalistic/historical sciences from the mathematical/predictive.[4]

The distinction between "Earth-directed" and "theory-directed" science is a bit overdrawn for the sake of argument. These perspectives are inevitably mixed in practice, yet merely drawing the distinction raises questions about current directions and emphases of the Earth sciences. In his discussion of the oral presentation of this essay at the 1998 annual meeting of The Geological Society of America, geophysicist Thomas Jordan observed that a new kind of empiricism is emerging at the interface between geology and physics. Jordan believes that this new science of "geosystems" is distinct from both the reductionism of modern physics and what he calls "the hermeneutics of classical geology."[5] In contrast to the methods of modern physics, it is not possible to subsume all the messy complexities of geosystems under fundamental principles of sublime beauty and symmetry. Instead, many general principles apply, including fashionable notions of fractals, chaos theory, and self-organized criticality. Geosystems include so many contingent elements that overarching principles can only apply to specific cases, and then only approximately. To this one might add the view of physicist John Ziman in his 1978 book *Reliable Knowledge: An Exploration for the Grounds for Belief in Science*. Physics, according to Ziman, concerns those aspects of reality that are amenable to mathematical analysis. It thus follows that the more geology becomes amenable to mathematics, the more it becomes a branch of physics.

In the introduction to his 1914 book *Igneous Rocks and Their Origins*, Harvard geologist Reginald Daly echoed many of the themes present in the current debate over the scientific nature of geology. "Geology," he noted, "has been charged with failure to measure up to the intellectual standard of the so-called 'exact' sciences." Daly ascribed this reproach to overzealous emphasis on the power of the experimental method by logicians of science. "Now that the intoxication of early, magnificent success in the use of experiment is succeeded by more sober second thought," Daly wrote, "it has become clearer that this method of research is only one of several that are quite essential and are of coordinate value in scientific thought." For Daly, the experimental methods of the "exact sciences" had both advantages and disadvantages: "Their mathematics is precise; their premises are not."

Daly would agree with Jordan's call for quantification. Daly observed, "Though not so tinctured with mathematics, geology…is 'exact' in the sense that a countless number of its observations are quantitative, with limits of error so small as to permit absolutely rigorous deduction." Quantification was not the critical issue in 1914, and it is not the key issue as we enter the twenty-first century. The important issue for Daly, and the one that most troubles me, lies not in the so-called systems of nature but in the systems of scientific thought. Daly wrote, "At bottom each 'exact' science is, and must be speculative, and its chief tool of research, too rarely used with both courage and judgment, is the regulated imagination." Daly concluded, "What geology, like every other science, needs to-day is a frank recognition that imaginative thought is not dangerous to science but is the life blood of science…. Science is drowning in facts. It can only be rescued by the growth of systems of thought…. The 'facts' of to-day are the hypotheses of yesterday."

In his brief essay, Daly argued for methods that contrast with those of prominent physicists like Henri Poincaré. Poincaré had observed in *The Foundations of Science* (1913) that science is built of facts as a house is of bricks. Poincaré echoed Wittgenstein's *Tractatus*, claiming that theory or method serves to assemble the facts of the world just as the thought and action of the home builder assembles the bricks. Daly clearly disagreed with this conventionalist view of science. Daly stated, "Science is built of a long succession of mistakes. Their recognition has meant progress. Progress, indefinitely more rapid, will be possible when men of science have more generally lost their fear of making mistakes in using the uttermost of their powers of correlation and deduction." Science is not built of facts, but of processes of thought.

Jordan views the science of "geosystems" as an iterated cycle of data gathering and analysis, predictive model formulation, hypothesis testing, and model improvement. This is the method of controlled experimentation, without a controlled laboratory. The larger part of Earth is inaccessible to controlled experimentation. The method must be applied to a "geosystem" of our definition. This is a trial-and-error construction from facts in which predictive modeling provides the principal tool of the system builder. Models generate predictions that are then compared to data collected from the real world. This strategy follows the current trend of empowering theory with computational efficiency. Predictive computer modeling has revolutionized the ability of scientists and engineers to consolidate theoretical knowledge into convenient conceptual packages that can be used to simulate the behavior of "systems" over ranges of conditions and for processes presumed to operate in the real world. Because of the increased ease in performing such modeling, the apparent rigor of the methodology, and utility of the results, predictive computer modeling has become a principal operating tool of modern Earth science.

Perhaps it would be well to define some terms at this point. A theory is an organized system of statements about the world. The organization involves various assumptions, including rules of logic and axiomatic principles. A theory both applies to a system, and is a system itself. The term "system" refers to some set of interrelated or interconnected elements. These elements are human abstractions, deemed to be important or essential interrelations or interconnections. The "true" system of all "real" interconnections and interactions is an intellectual ideal, the realization of which is ultimately either unknowable or untestable, because of limitations on our ability to access all the complex connections of the real world.

Models are human constructs that simplify reality. Models embody elements abstracted from the world by their inventors, such that these elements represent only essential workings of that world, not the whole of its real complexity. Thus, the model system is both fully known and testable. The advantage of this model system is that, through its relative ease of manipulation, the modeler can be led to potential understanding about the more complex real world that the model has been presumed to represent. The disadvantage is that the issue of representation is far more complex and logically problematic than most modelers realize.

In applied/engineering modeling, the best available theory is incorporated into models to achieve accurate representation (simulation) of the system of interest. In the engineering case, this is a system predefined by exact criteria for design, management, or control. It is necessary to obtain facts through direct measurements on this system, facts that we call "data." Here, the data are used to make initial estimates of system parameters and to help the selection of appropriate model parameters that reduce uncertainties in the prediction of processes. In other words, data serve to calibrate the models. Thus, engineering models apply to specific, predefined criteria. They are valid only for the problem domain to which those criteria apply.

To be scientific, a theory must relate to the empirical world—that is, to facts and observations, measured or sensed from the world of experience. While engineering models are calibrated to perform as practical tools, geophysical models function to advance scientific understanding, and ultimately to achieve new theories.

The simple experimental approach to the testing of models for complex environmental systems encounters a major logical difficulty, one that was elaborated by Wittgenstein's successors in the philosophy of language. One way to state the problem is as follows: "[M]odel verification is only possible in closed systems in which all the components of the system are established independently and known to be correct."[6] It is not sufficient that model predictions match observational data. Such a result can be achieved through numerous combinations of model assumptions or parameter adjustments. Indeed, a famous principle of logic, the Duhem-Quine thesis, explicitly disallows such model verification. Simply put, the Duhem-Quine thesis precludes conclusive theory (model) falsification because any predictive clash with observation can always be blamed not on the theory itself, but rather on the maze of assumptions (some not explicitly stated) in the test situation.

For the complex world of environmental processes, models cannot be strictly verified. This is true for several reasons: Real-world systems are not closed, the model predictions are nonunique (the Duhem-Quine thesis), model input parameters and assumptions are almost always poorly known, and scaling issues for nonadditive properties are also poorly known. In addition, the data themselves pose problems, and this latter issue is central to geology. Can we treat geological data as exact substitutes for some natural phenomena, or should we consider these data themselves to be inference-laden signifiers of those phenomena? If it is the real world that is our primary interest, rather than various theories about it, how do we deal with our limited access to natural phenomena? For much of our data consists of *signifiers* of phenomena, rather than being a substitute or proxy for natural phenomena. For instance, the geochemical data taken from ice cores are themselves inference-laden signifiers.

Inference-laden signification offers an alternative to the metaphysical presumption of theories mirroring nature and data substituting for phenomena. This alternative involves a philosophical point concerning a scientist's view of the world. The Earth-directed point of view is *semiotic* in that it understands data as part of a complex interpretive structure mediated and sustained by signs. In semiotics, signs stand for something (the object) in relation to something else (the interpretant). The signs of geology are indexical signs, in which they relate to objects in terms of causation. Thus, various causal indicators (indexical signs) point to causal processes as their objects—for instance, the sediments that point toward extreme floods. Although the natural indices found in the field, including various sediments and landforms, are implicitly semiotic in themselves, their interpretant function is realized via thoughts triggered in human investigators. Those thoughts, in turn, become signs leading to semiotic activity continuous from the natural world to the world of human thought. It is through this semiosis, or action of signs, that the Earth converses with the geologist.[7] To understand how this occurs will require an analogy.

The notion of data as indexical signs, rather than as proxies used in theory testing, can be made clear by considering a detective at a murder scene. The detective finds a small hole in the wall with an embedded lead particle. This observation might serve as the test of the theory (model) that a gun was used in the crime, firing the bullet, making the hole, and so forth. Conceptually, the presumed bullet hole serves as a proxy for the gun, a murder weapon that is missing. Of course, other theories can explain these circumstances. Perhaps the gun firing the bullet is not related to the crime. In the semiotic viewpoint, however, the crime theory is not the immediate concern. Instead, the detective is a student of bullet holes, much as Sherlock Holmes was a student of exotic tobacco. It is a rather secure notion that the hole and the embedded projectile provide an index of a process: gunfire. The detective studies this sign (clue) and combines it with other clues, developing a web of interconnecting clues (signs). Eventually, a narrative connecting these clues emerges as the working solution to the crime. The key element of the investigation is not the testing of theories. Rather, it is the binding together of facts and the overall consistency and coherence of the working solution (hypothesis) with the complex, developing web of interconnected clues or signs.

Although the detective story illustrates semiosis,[8] it still has limitations in regard to geology. Every clue (sign) is filled with potential for understanding the world through its causal connections to the world. This potential extends beyond any initially defined crime scene ("system") to connections that lead the investigator to a new understanding. It is these connections of semiosis that are most productive for the systems of geological thought.

Galileo Galilei once observed that the book of the universe is written in the symbolic language of mathematics. To read the book of the universe, one must learn this language. The Earth's geological book also has its mathematical language, but this symbolic language is not the only one worthy of inquiry. The language of indices, signs directly representing causative processes, constitutes another text for reading. This text cannot be fully understood in symbolic terms, which involve language imposed by our own conventions. The indexical language of geology is learned from nature itself. Landforms and sediments emplaced by past processes constitute the sign language to be interpreted. By adopting a semiotic point of view, the geologist is drawn into a process of inquiry that leads to understanding nature's reality *prior* to the definition of systems appropriate for explanation through model simulation. This process of inquiry is only obliquely addressed by a viewpoint that treats nature's signs as proxies for our conceptualizations, and then uses these proxies to test the validity of the model simulations. Not only is such model validation/verification logically flawed for the real (open) systems of nature, but it can stifle the spirit of inquiry associated with the semiotic viewpoint. Here is where I see a defect in Jordan's science of geosystems. The "geosystems" approach limits one to the world of model predictions and objective data gathering.

The Earth can speak to us semiotically just as great books speak to us. We can converse in these inquiries if only we possess the imagination to do so. As Einstein once observed, "Imagination is more important than knowledge."

It is the ingrained habit of modern scientists to view Earth as a totally mindless, inanimate object. This habit is so developed that, without reflection upon the matter, many scientists will presume the existence of a razor-sharp separation between (a) an Earth of pure objective facts and (b) knowing qualities of mind ingrained in scientists. Moreover, it is even presumed that this separation constitutes some sort of ontological necessity of nature. The latter claim is one of metaphysics, not science. There is no way to "test" by controlled experimentation whether the mind/Earth separation is a necessary condition of nature. Indeed, there is no way even to test whether the distinction is so hard and fast as presumed.

Probably most scientists would agree with Sir Isaac Newton's dictum: "Physicist beware metaphysics!" Rather than claiming an ontological basis, scientists might well consider the Earth/mind distinction to be a necessity for reasoning. One cannot be logically certain about knowledge of the Earth unless one makes this distinction. This is an epistemological claim, and volumes of logical philosophy have been devoted to analyzing its merits. Nevertheless, this is also a metaphysical doctrine in that it claims, *a priori*, a method for reasoning. Yet another view is not to prejudge whether a particular method works in generating productive scientific discoveries, the merits of which cannot be anticipated *a priori*. Though scientists may argue endlessly over methods to be used, they tend to converge to agreement *a posteriori*—that is, upon the accordance of discovered relationships with the actual operations of nature.

If one views nature as full of semiotic potential, with its signs continuous from its web of causations to the beliefs of scientific minds, then one has a method of demonstrated fruitfulness in geological inquiry. This method is a conversation to the degree that hypotheses constitute questions, rather than propositions to be tested. Moreover, data constitute directed replies, or signs of continuous thought, rather than numerical symbols to be mirrored by theoretical abstraction. Certainly the metaphysics of conversing with nature is no more extreme than that of defining nature's systems. It would seem that the contrast is between our language naming the facts of the world that it is presumed to mirror, and a language of which we are a part and that extends from the world to us. The latter has been termed a "conversation," but, like the naming of facts, it must ultimately demonstrate its productivity in use. Perhaps such utility is more fundamental to science than is the metaphysics of language, representation, and verification/falsification.

I wish to conclude by returning to the work of Ludwig Wittgenstein. The rigorous logical analysis of his *Tractatus Logico—Philosophicus* led ultimately to an enigmatic, disturbing conclusion: "Whereof one cannot speak, thereof one must remain silent." If we limit ourselves to the language of what we can say of the world, then we are closed in a box of self-consistent logical clarity, but one of lonely detachment from what is outside the box. In his mature career, propelled by the fame of the *Tractatus*, Wittgenstein came to reject nearly all that he had so rigorously advocated. In the *Tractatus* he had presumed that the meaning of words in language was limited to their naming of facts. Ultimately, this view of meaning leads to the same kind of despair expressed by Umberto Eco in the last sentence of his novel *The Name of the Rose*: "We have only names."

Wittgenstein resolved his language dilemma by concluding that meaning for language is found in its use. Small children do not develop language by logical analysis. It grows with them through use. Rules eventually develop until the language functions like a game. For the early Wittgenstein of the *Tractatus*, philosophy was the process of bringing clarity to language, in essence to make the best possible statements about the world. For the later Wittgenstein, the task of philosophy was to help us get past this boxed-in way of viewing the world. Wittgenstein wanted to draw attention to "a bewitchment of our intelligence by means of language." He likened the language trap to a fly caught in a fly bottle. He claimed that his aim in philosophy was "to show the fly the way out of the fly bottle."

I suggest that geology has developed its own way out of the fly bottle. It does not accept the restrictive role of a symbolic language. It embellishes what we can say about the world with what the world signifies to us. Moreover, this signification is a very rich and precious resource, and it needs to be understood as part of the system of scientific thought.

To preserve a continually habitable planet through an uncertain future of global environmental change, science will need to develop the best possible mathematical models to explain that future. However, just as those models will need to embody the best that scientists can say about the real world, it will also be necessary to explore the real experience of that world for the best that it can say to us. The challenge to geology is for its practitioners to devote just as much effort to exploring the real world of Earth experience as they devote to idealizing the abstract world of geosystems. By preserving this essential tension, science will enhance the quality of human existence.

NOTES

1. As quoted in Petersen, "The philosophy of Niels Bohr," p. 305.
2. William Whewell's debate with Charles Lyell over the logical basis of geology is described by Baker, "Catastrophism and uniformitarianism: Logical roots and current relevance in geology," pp. 171–182.
3. Cloos, *Conversation With the Earth*.
4. The logic of hypotheses in the Earth sciences is described by Baker, "Hypotheses and geomorphological reasoning," pp. 57–85.
5. Jordan, "Response by Thomas H. Jordan," p. 25–26.
6. Oreskes and others, "Verification, validation and confirmation of numerical models in the Earth sciences," pp. 611–646.
7. The role of semiosis in geology is described by Baker, "Geosemiosis," pp. 633–646.
8. The relationship of semiotics to the logic of the detective is described by Eco and Sebeok, *The Sign of Three: Dupin, Holmes, Peirce*.

REFERENCES

Baker, Victor R. "Hypotheses and geomorphological reasoning." In *The Scientific Nature of Geomorphology*, ed. B.L. Rhoads and C.E. Thorn, pp. 57–85. New York: Wiley, 1996.

Baker, Victor R. "Catastrophism and uniformitarianism: Logical roots and current relevance in geology." In *Lyell: The Past is the Key to the Present*, ed. D.J. Blundell and A.C. Scott, pp. 171–182. London: The Geological Society Special Publ. 143, 1998.

Baker, Victor R. "Geosemiosis." *Geological Society of America Bulletin*, 111 (1999), pp. 633–646.

Cloos, Hans. *Conversation With the Earth*. New York: Knopf, 1953.

Daly, Reginald A. *Igneous Rocks and Their Origins*. New York, McGraw-Hill, 1914.

Eco, Umberto, and Sebeok, Thomas. *The Sign of Three: Dupin, Holmes, Peirce*. Bloomington: Indiana University Press, 1988.

Jordan, Thomas H. "Response by Thomas H. Jordan." *GSA Today*, 9, no. 3 (1999), pp. 25–26.

Oreskes, N., Shrader-Frechette, K., and Berlitz, K. "Verification, validation and confirmation of numerical models in the Earth sciences." *Science*, 263 (1994), pp. 611–646.

Petersen, Aage. "The philosophy of Neils Bohr." In *Neils Bohr: A Centenary Volume*, ed. A.P. French and P.J. Kennedy, pp. 299–310. Cambridge, MA: Harvard University Press, 1985.

Whewell, William. *History of the Inductive Sciences*. London: Cass, 1837.

Wittgenstein, Ludwig. *Tractatus Logico-Philosophicus*. New York: Humanities Press, 1961.

Ziman, John. *Reliable Knowledge: An Exploration for the Grounds for Belief in Science*. Cambridge: Cambridge University Press, 1978.

2

Reading the Riddle of Nuclear Waste: Idealized Geological Models and Positivist Epistemology

Kristin Shrader-Frechette

Kristin Shrader-Frechette is De Crane Chair, Professor of Philosophy, and Concurrent Professor of Biology at the University of Notre Dame. Since the 1970s she has seen the problem of risk, as it arises with the development and utilization of modern technologies (especially nuclear technologies) as a central but largely neglected issue for philosophy. In several books—including Risk and Rationality: Philosophical Foundations for Populist Reforms *(Berkeley: University of California Press, 1991), and* Burying Uncertainty: Risk and the Case Against Geological Disposal of Nuclear Waste *(Berkeley: University of California Press, 1993)—she has systematically sought to remedy this oversight.*

In this essay, Shrader-Frechette explores the nature and limits of geological models for guiding public decision making. Too often, both scientists and members of the community fail to distinguish between well-established scientific fact, and the opinions of scientists. Schrader-Frechette identifies problems with the logical pre-suppositions of models themselves, the ambiguous nature of the data used in the models, and with the over-reliance invested in them by both scientists and the public. Basing her argument on the problems surrounding the long-term disposal of nuclear waste, Schrader-Frechette demonstrates that the models used rely upon poorly understood processes, methodological value judgments, and presuppositions that cannot be demonstrated. Given the uncertainties involved, decision making must be an open, democratic process where citizens and scientists engage in a dialogue about the nature of the choices faced.

Nuclear waste is similar to the Ring in J.R.R. Tolkien's *The Lord of the Rings* (1965, pp. 349–350). The Ring gave mastery over every living creature, but it was created by an evil power. As a result, it inevitably corrupted anyone who attempted to use it. The hobbits, who held the Ring, had to decide how to handle it. The character Erestor formulated the problem: "There are but two courses, as Glorfindel already has declared: to hide the Ring forever, or to unmake it. But both are beyond our power. Who will read this riddle for us?"

2.1 OVERVIEW

Scientists all over the world have been attempting to read the riddle of dangerous radionuclides. Microbiologists and molecular biologists have been following the latter course, to "unmake" nuclear waste. They are pursuing goals such as genetic manipulation of the bacterium *Deinococcus radiodurans*, a microbe that can both shield its DNA from ionizing radiation a thousand times greater than what would kill humans and that can repair genetic damage from radiation. Although no known bacterium can actually metabolize uranium or other metals into harmless substances, some microbes do have genes encoding proteins that immobilize metals with which they come in contact. (Travis 1998). Until some sort of "unmaking" of nuclear waste works, and it may not, scientists will continue to pursue the former course mentioned by Erestor: hiding it forever. But "hiding" radioactive materials requires that scientists have precise knowledge of the hydrology and geology of the burial site. This in turn requires rigorous hydrological and geological models able to predict whether radionuclides might migrate out of deep burial grounds.

How good are the hydrogeological models that predict "radwaste" migration? The standard wisdom is that researchers ought to use the best scientific models until better ones become available. For tracking radioactive waste (radwaste) through groundwater, porous media models probably are the best available, and hydrogeologists have used them extensively. When public health and safety are at issue, however, the best models may not be good enough. When used in real-life situations having practical consequences for human health and welfare, current geological models may be so imprecise and idealized that they risk catastrophic public consequences. The essay argues that (1) scientists often use idealized geological models inappropriately, in part because the scientists confuse facts and values. Using examples from a recent National Academy of Sciences report on standards for storing nuclear waste, I show that (2) reliance on idealized geological models and expert guesses is dangerous because it often is confused with confirmed science. Because of this confusion, I suggest that (3) scientists and policymakers tend to ignore the concerns of stakeholders who question the use of idealized models in potentially catastrophic situations. I close by showing that (4) there are a number of criteria for deciding when and how to use idealized geological models in practical policy situations.

2.2 THE GEOLOGICAL PROBLEM AND THE POLICY STAKES

In the absence of geological proof that people can successfully store radwaste, in perpetuity, the technical problems associated with it are forcing policymakers to take a great gamble. This is a gamble that future descendants will not breach nuclear repositories through war, terrorism, or drilling for minerals; a gamble that water and heat will not combine to create nuclear reactors in underground waste, as already happened in the former Soviet Union; and a gamble that ice sheets, volcanism, seismic activity, and geological folding will not uncover the radwastes.

Scientists had not been worried earlier about these gambles because, for the first 35 years of commercial nuclear fission, they were saying that safely isolating the wastes would be easy, once they set their minds to the task. They were like contractors who built houses without toilets and then alleged that constructing the toilets would be easy.

In Section 801 of the Energy Policy Act of 1992 (PL, 102–486), the U.S. Congress directed the Environmental Protection Agency (EPA) to promulgate standards to ensure protection of public health from high-level radwaste in a permanent geological repository that the government wants to build at Yucca Mountain, Nevada. This provision of the 1992 act requires the EPA to set standards to protect the health of individual members of the public. To assist the EPA in this standard-setting, Congress also asked the National Academy of Sciences to advise the agency on the technical bases for such standards. In August 1995, a committee of the Board on Radioactive Waste Disposal of the National Research Council (NRC) published *Technical Bases for Yucca Mountain Standards*, the advice for which the Congress asked.

Technical Bases is a landmark document that significantly advances understanding of both the science and the policy relevant to high-level radwaste disposal. It has many assets, especially its recommendations that compliance with the risk standard for radioactive waste be measured at the time of peak risk, whenever it occurs (National Research Council [NRC], 1995, pp. 2, 55–56, 67); its conclusion that there is no scientific basis for limiting health-and-safety concerns merely to 10,000 years (NRC, 1995, p. 56); and its important stance on protecting intergenerational equity. The Academy of Sciences' document is clear and straightforward about many important uncertainties in its recommendations about radioactive waste disposal, about site modeling, and about performance assessment generally (NRC, 1995, pp. 19–20). It also does an excellent job of emphasizing that there is no sharp dividing line between science and policy, and that there is a limited scientific basis for choosing one policy option over another (NRC, 1995, p. viii).

2.3 SCIENTISTS' CONTROVERSIAL USE OF IDEALIZED GEOLOGICAL MODELS

Despite all its assets, the Academy of Sciences' report expresses confidence in something that is quite questionable: estimating long-term (million-year) security of nuclear waste on the basis of idealized geological models. The problems lie both with the models themselves and with the apparent overconfidence invested in them. Consider two recent uses of geological models, one at Yucca Mountain, the proposed high-level waste-disposal site, and the other at Maxey Flats, an existing low-level waste facility. Although the sites are geographically diverse and intended for different materials, the problems of hydrogeological idealization are similar at both locations.

Yucca Mountain, Nevada, the proposed site of the world's first permanent repository for high-level radioactive waste, is located near Las Vegas in an area of high seismic and volcanic activity but low rainfall, approximately 13 inches annually. Since 1987, U.S. Department of Energy (DOE) contractors have spent more than \$3 billion to determine the area's feasibility for perpetual storage of radwaste and spent nuclear fuel. The DOE hopes to store the waste, at a temperature of 200°C, in single-walled, stainless-steel containers. To do so, it must be able to construct a plausible repository performance assessment, including groundwater behavior. But because of the long half-lives of the radionuclides being stored, the performance assessment must cover a million or more years into the future (NRC, 1995). In 1992, a 14-person DOE peer-group review team of hydrologists and geologists unanimously concluded that no specific predictions, over so long a time frame, were possible owing to massive uncertainties (Younker et al., 1992).

> It is the opinion of the panel that many aspects of site suitability are not well suited for quantitative risk assessment. In particular are predictions involving future geological activity, future value of mineral deposits and mineral occurrence models. Any projections of the rates of tectonic activity and volcanism, as well as natural resource occurrence and value, will be fraught with substantial uncertainties that cannot be quantified using standard statistical methods. (Younker et al. 1992, p. B-2; see Shrader-Frechette 1993, pp. 123–124, 152–153, 164–168, 175)

In August 1995, however, a National Research Council Committee of the Board on Radioactive Waste Management (NRC, 1995, p. 91) said that the uncertainties could be overcome and that it was possible to estimate reliably hydrogeological and repository performance over the next million years:

> We conclude that the probabilities and consequences of modifications generated by climate change, seismic activity, and volcanic eruptions at Yucca Mountain are sufficiently boundable so that these factors can be included in performance assessments that extend over periods on the order of about 10^6 years. (NRC, 1995, p. 91)
>
> Established procedures of risk analysis should enable the combination of the results of all repository system simulations into a single estimated risk to be compared with the standard. (Human intrusion is excluded from such a combination). (NRC, 1995, p. 69)
>
> Processes are sufficiently boundable that they can be included in performance assessments that extend over time frames…on the order of about 10^6 years. (NRC, 1995, p. 85)
>
> It is possible through careful examination of the geologic record to establish a chronological history of the activity over millions of years. Estimates of activity over similar periods into the future can be made by extrapolation from the past activity. (NRC, 1995, p. 93)

As the previous claims indicate, some of the geological questions raised by the National Academy of Sciences' report include the following:

(1) If the committee believes that future societal events cannot be predicted (NRC, 1995, p. 96), but if future societal events could influence "geological engineering factors," then how are geological engineering factors susceptible to realistic, long-term estimation, as the committee claims?

(2) Given data gaps and therefore massive uncertainties and problems with verification and validation of performance assessments, how can million-year estimates of geological events be reliable?

(3) Are claims about repository compliance and million-year geological estimates matters of expert opinion of science?

(4) Having claimed that there are serious uncertainties about $^{14}CO_2$ exposures (NRC, 1995, pp. 87–88); about nonuniform radionuclide distributions (pp. 88–89); about

fracture flow, especially in the unsaturated zone (pp. 88–90); and about "several glacial periods" during the million years of the repository (p. 97), how is the committee able to affirm confidence in the million-year geological estimates already mentioned?

More generally, who is right about using the idealized models at Yucca Mountain—the National Academy of Sciences' Committee (NRC, 1995) or the DOE peer review committee (Younker et al., 1992)? Before answering this question, consider another case, one in which hydrogeological predictions (made from idealized models) turned out to be wrong.

Maxey Flats, Kentucky, is the location of the world's largest (in curies) commercial low-level radioactive waste dump (Powell, 1991, p. 31; Eschwege, 1976, p. 4; see also Haney, 1994; and Meyer, 1976, p. 1). Located in the poverty-ridden Appalachian foothills, on land purchased secretly, Maxey Flats contains more plutonium than any other commercial facility and is known as "the world's worst nuclear dump" (Powell, 1991, p. 31). The corporate lease-holder, Nuclear Engineering Company ((NECO), name later changed to US Ecology), opened the site in 1963 and predicted that it would take plutonium 24,000 years to migrate one-half inch at the site (Neel, 1976, p. 245; Hopkins, 1962; see also Weiss and Columbo, 1980, p. 5; Shrader-Frechette, 1993, p. 5). NECO was wrong, by six orders of magnitude. Less than a decade after Maxey Flats opened, plutonium—with a half-life of more than 25,000 years—had migrated two miles off-site (Meyer, 1975, p. 9; Hartnett, 1994; Bedinger and Stevens, 1990, pp. 34–35; Powell, 1991; see Hyland et al., 1980, 1982). The taxpayers of Kentucky, among the poorest in the nation, are now contributing approximately $1 million annually to maintain the closed facility in perpetuity, but the radwaste continues to migrate off-site (Hyland et al., 1982, p. 36). Scientists say it is a potential public health hazard (Hartnett, 1994; Powell, 1991, p. 32; Clark and Wilson, 1974, p. 14; see Haney, 1994), and Maxey Flats is on the Environmental Protection Agency's (EPA) list of sites requiring priority action. It is the only radwaste facility on the Superfund list (Hartnett, 1994, p. 351; Bedinger and Stevens, 1990, p. 32).

2.4 PROBLEMS WITH USING IDEALIZED HYDROGEOLOGICAL MODELS

Apart from the political questions raised by Yucca Mountain and Maxey Flats, there are also controversial scientific issues: How could a National Research Council committee argue that million-year estimates of Yucca Mountain hydrogeology were plausible? How could the scientists at Maxey Flats be wrong, by six orders of magnitude, in their predictions about waste migration? Both situations raise the question whether to accept or reject a given scientific model—for example, of groundwater migration—when the situation is empirically underdetermined. How does one frame such questions of model choice, especially when they have important policy consequences?

On the one hand, as already mentioned, philosophers of science argue that it makes no sense to reject a model until a better one is available. Laudan (1977, pp. 26–44) suggests, for example, that only if a competitor model or theory is able to solve problems better, or improve upon some flaw, does the flaw pose a cognitive threat to the original theory. On the other hand, moral philosophers claim that, even if a better model is not available, some models ought to be rejected. They reason that even the best model may not be good enough, especially if it could lead to undesirable public policy consequences (see Sagoff, 1985).

Scientists and philosophers of science might argue for using flawed models at Yucca Mountain and Maxey Flats, provided they are the best available. Ethicists might argue for rejecting both models, even if they were the best available. Which criteria for model choice, in policy-relevant, underdetermined scientific situations, might be more desirable? Examining the Yucca Mountain and Maxey Flats cases might provide some insights about model choice. Both situations require that people know something about hydrogeological models, because the main guarantor of the immobility of radioactive waste is the site hydrology and geology (in the Yucca Mountain case, volcanic tuff; in the Maxey Flats case, various shales). If the wastes migrate, it would be through leaching and water transport.

Most of the models used in hydrogeology are simulation models, mathematical expressions that attempt to describe or explain some physical phenomenon, such as dispersion of a subsurface liquid throughout the underlying geological strata. A major objective of the models is to account for the time and space variability of various hydrogeological components. Before most hydrogeological models can be used for prediction, detection, or sensitivity analyses, however, they must be calibrated. The Achilles' heel of the entire calibration and modeling process is the fact that no hydrogeological model, either in terms of its system assumptions or in terms of particular values for its parameters, can be validated. This is because of the inadequacy and inaccuracy of the data, which are supposed to act as a check on the reliability of the model (Fleming, 1975, pp. 237, 257). At Maxey Flats, the period of interest is tens of thousands of years, four orders of magnitude longer than any period of observation (Carey, Lyverse, and Hupp, 1990, p. 32; see Lyverse, 1990). In the Yucca Mountain case, the period of interest is millions of years, but hydrogeological data exist for only about 10 years. Spatial and temporal heterogeneities, at both sites, also make most generalizations unreliable, in part because infrequent events at the disposal sites can do as much damage as thousands of years of normal geological processes (Carey et al., 1990, pp. 34–35). In the face of such empirical difficulties, computer scientists and mathematical modelers rely on benchmarking (that is, on the consistency of results in two or more formal computer models), rather than on genuine empirical confirmation as the basis for model choice (Shrader-Frechette, 1993, pp. 50, 147; Oreskes et al. 1994).

Hydrogeologists rely on benchmarking in part because of problems with data collection. The heart of these problems is that only by unearthing an area, and rendering it unsuitable for storing hazardous wastes, is it possible to know exactly the characteristics of the underlying hydrology and geology. Yet only by not disturbing an area, and hence not discovering its precise hydrogeology, is it possible to make it secure for storing nuclear wastes like those at Yucca Mountain and Maxey Flats. Data collection also is problematic because site heterogeneities and fracturing at both Yucca Mountain and Maxey Flats undercut representative sampling.

The chief alternative to using poor measured values for parameters such as "soil moisture" is model calibration, manipulating a specific model to reproduce the field response within some range of accuracy. As already mentioned, because such models cannot be calibrated on the basis of long-term measurement, their accuracy cannot be fixed by establishing a criterion of fit between the theoretical, simulated responses of the model and the actual responses of real-world phenomena. Moreover, ignorance of real responses, over the long term, is precisely what led to use of models in the first place (see Shrader-Frechette, 1993, pp. 146ff; Fleming, 1975, pp. 237, 257). The theoretical models are necessary because knowing initial and boundary conditions is not sufficient for precise, long-term prediction of real systems. Many hydrogeological situations are so complex, heterogeneous, and unique that no others are similar, and replication is impossible. The "uniqueness problem" would not occur in situations that are hydrogeologically uniform—for example, completely sandy soils with little or no rainfall, rather than (as at Maxey Flats) bedding planes of shales, erratically fractured, in an area of high rainfall (Wilson and Lyons, 1991). In short, the hydrogeological models used at Maxey Flats and at Yucca Mountain cannot be validated because of difficulties of data gathering and site heterogeneity.

In addition to data gathering, a second problem is that hydrogeological models rely on field conditions that often do not obtain, as is the case both at Maxey Flats and at Yucca Mountain. Whether hydrogeologists rely on physical-scale models, like porous-media models—on the analogy between water flow and laminar flow of fluids, heat, and electricity; on the similarity between Ohm's law and the Darcy equation, $v = -K\,dh/dx$; or on the computational procedure of dividing an aquifer into a grid and analyzing the flows associated within a single zone in terms of partial differential equations (see Todd, 1980, pp. 384–400)—all these groundwater-flow models presuppose field conditions that rarely obtain (see Carey et al., 1990, pp. 32ff).

For example, most of the models depend upon Darcy's law, which applies to laminar flow (flow through thin layers) in porous media, but not to turbulent flow, such as can be found in Maxey Flats' limestone rocks, which contain large underground openings. Other groundwater-flow models require idealized, oversimplified conditions—such as that the aquifer in question is homogeneous and isotropic (has the same properties, e.g., conductivity, in all directions); has impermeable boundaries at the side and bottom; or has a flow region that is rectangular (see Todd, 1980, pp. 67–68, 93, 101, 342)—that neither Maxey Flats nor Yucca Mountain meets (Wilson and Lyons, 1991; Zehner, 1981, pp. 1, 3, 73–77, 112, 119–126, 132–134, 153, 157, 175, 181; see also Werner, 1980; Shrader-Frechette, 1993).

At Maxey Flats, for example, the most active groundwater flow is in the upper part of the hill (not the lower) and through fractures (Bedinger and Stevens, 1990, pp. 33–34). Indeed, at both the Maxey Flats and the Yucca Mountain sites, fracturing is extensive. Because there are no models for fractured sites, like these two, only rough estimates of groundwater flow are possible (Bedinger and Stevens, 1990, p. 35; Trask and Stevens, 1991, p. 47; Schoeller, 1974, p. 4; Todd, 1980, p. 59; Zehner, 1981; Werner, 1980). The optimists about hydrogeological prediction at both sites take no account of fractures and fracture flow, whereas the predictive pessimists argue that fractures preclude accurate modeling (see Shrader-Frechette, 1993, pp. 61–66).

Key problems with groundwater-flow models (lack of validation and contraindicated site conditions) require scientists to exercise considerable judgment in framing hydrogeological questions and choosing predictive models. Often they frame site questions in ways that ignore the limits of their models and then draw dramatic conclusions about the relative hydrogeological suitability of different proposed waste sites. As a result, hydrogeologists often have a false sense of security about the respective safety of nuclear waste at different locations. In assessing the Maxey Flats site, EPA geologists specifically noted that using porous-media models could result in grossly underestimating the groundwater velocities (Papadopulos and Winograd, 1974, p. 34). Nevada geologists made the same point in criticizing DOE models of Yucca Mountain (see Shrader-Frechette, 1993, pp. 40–41). Their criticisms suggest that, whenever scientists cannot specify the empirical fit of idealized models, they ought to question the epistemological rule of using the "best available" model.

A third difficulty (in addition to using models that cannot be validated and that do not fit site conditions) with using the best hydrogeological models at waste sites is that they are typically not testable and are rarely falsifiable. Problems associated with testing groundwater-model predictions by means of tracer methods provide a good illustration of this point. Apart from the difficulties with the tracers themselves (and the most common is tritium, which confuses results because it is commonly released from waste sites), groundwater movement is often slow. Because of this slowness, tracers are little help in testing million-year Yucca Mountain predictions (see Wilson and Lyons 1991). Using tracers also requires one to discount drilling effects on natural flow and dispersion of the water. Unless hydrogeologists can guess the flow direction, the tracer may miss the downstream hole entirely. Multiple sampling holes can help solve this problem, but they disrupt the hydrogeological environment that one is attempting to monitor. Also, because of diffusion, tests must be conducted over short distances to have detectable concentrations at the downstream well.

Nevertheless, it is difficult to determine a representative time of arrival (Todd, 1980, p. 75; Linsley, Kohler, and Paulhus, 1975, p. 203). In situations of fractured media, like Maxey Flats and Yucca Mountain, the patterns of flow are quite erratic, so that determining a good tracer checkpoint is difficult. Ian Walker, the scientist who performed the siting work for Maxey Flats, explicitly noted this problem:

> At Maxey Flats it would be impossible to predict the path of a ribbon of contaminant in order to locate monitoring wells as the contaminated water would tend to undulate to seek permeable joints in the shale and to avoid the impervious solid matrix. (Walker, 1962, p. 2; see Trask and Stevens, 1991, p. 47; Wilson and Lyons, 1991)

A fourth, and more serious, difficulty with hydrogeological models is that scientists often do not agree on what would confirm them. Problems at Maxey Flats (such as not obtaining gamma measurements, not taking account of significant lateral movement through the shale, not sampling at some spots deemed crucial by U.S. Geological Survey and EPA scientists, not checking for particular radioactive isotopes, and not knowing where to monitor for tracers) indicate that the scientists did not agree on what parameters were necessary for legitimate testing of their models (see Trask and Stevens, 1991, p. 47; Papadopulos and Winograd, 1974, pp. 9–16, 35–36, 40–42; Blanchard, Montgomery, and Kolde, 1977, pp. 58, 68–69, 74). Similar disagreement (over what was necessary to test hydrogeological models) occurred at Yucca Mountain (Shrader-Frechette, 1993, pp. 53–56).

2.5 SCIENTIFIC OPINION, SCIENTIFIC CONFIRMATION, AND POSITIVIST EPISTEMOLOGY

Because hydrogeologists use models (for evaluating potential nuclear-waste sites) that can be neither validated nor tested, because they use the models in situations that do not meet model specifications, and because scientists do not agree on the criteria for model confirmation, much of what passes for scientific *findings* about geological suitability of waste sites is really *scientific opinion*. In other words,

(1) When a hydrogeological site is so *heterogeneous* that uniform values of key parameters are neither representative of the site nor consistent with model *requirements*;

(2) when *validation* is impossible because the empirical fit of idealized models is unknown;

(3) when the predictions generated by the model admit of no realistic *testing*; and

(4) when scientists do not agree on the parameters relevant to reliable model prediction and *confirmation*,

then it is questionable whether scientists are doing more than merely giving their opinions—or educated guesses—about the hydrogeological suitability of a particular waste site. The fact that the conclusions of the National Academy of Sciences are educated guesses can explain why they contradict the opinions of the hydrogeologists who were DOE peer reviewers for Yucca Mountain. The fact that the Maxey Flats predictions were scientific opinions can explain why off-site migration proved them wrong, by six orders of magnitude.

In attributing more confidence to their scientific opinions than the facts warranted, both the members of the National Academy committee and the Maxey Flats assessors appear to have fallen victim to naive positivism. Naive positivism is the belief that science can be value-free and completely objective. The naive positivists forget that all science is laden with methodological value judgments—about how to interpret data, how much data is necessary, how many samples are required, which curve best fits the data, and so on—and that, because of these value judgments, science is never wholly objective or value free. According to Mayo (1997), metascientists recognize the value judgments inherent in (even the best) science, and they critically assess these value judgments to determine which are more reliable.

Those who use the naive-positivist epistemology, however, fail to recognize the value judgments, and as a consequence they fall victim to them. They claim more certainty for their scientific conclusions than their models and methods warrant. Such a positivist error appears to be at the heart of the mistakes about both Yucca Mountain and Maxey Flats. Hydrogeologists failed to take into account methodological value judgments about the role of fracture flow, the uniformity of the subsurface water, and so on. Once one takes account of these methodological value judgments, it is likely that one will have a clearer idea of the reliability of one's scientific opinions and conclusions.

This brief overview of the controversies and contradictions inherent in using idealized hydrogeological models suggests at least two conclusions. One is that scientists ought to be careful to present scientific opinions as scientific opinions and not to claim more status for them than they have. They ought to avoid naive-positivist epistemology and instead to practice metascience—that is, critical evaluation of their science. A second conclusion is that scientists ought to devise ways to strengthen the scientific foundations of their claims about hydrogeological site suitability. One way to gain this strength, in addition to doing metascience, might be to encourage policymakers and scientists to prefer a *better modeled geological site* (such as sand) to a *better site* (such as shale), because at least scientists would know their margin of error. At least they could design repositories so as to take account of known problems. The reasoning behind choosing more predictable but less geologically secure sites is that the devil you know may be better than the devil you do not know. If this reasoning is correct, then dry, uninhabited sites (that are hard to model/predict) may be less desirable for nuclear waste than wetter, habited sites (that are easier to model/predict). At least predictions about the latter sites would be more susceptible to scientific confirmation and less susceptible both to scientific controversy and falsification.

2.6 SCIENTIFIC OPINION AND STAKEHOLDER REPRESENTATION

Of course, if scientists and policymakers prefer to consider dry, uninhabited sites that are harder to model/predict over wetter, more populous nuclear waste sites that are easier to model/predict, then it follows that siting decisions will need to incorporate more consideration of ethical and policy issues. The reasoning behind this conclusion is that scientists have the right, because of their expertise, to speak about largely scientific issues. They do not have the right to speak about the acceptability of specific responses to ethical and policy issues (such as how to behave under situations of site/modeling uncertainty). Because such ethical and policy issues affect other citizens, and involve risks to public welfare, they are decisions that the public ought to make. Because whether to take such risks is both an ethical and a policy decision, stakeholders (those affected) need to be involved in resolving issues about how to store radioactive waste when no sites can be confirmed as scientifically suitable. In other words, for policy-relevant situations that are empirically underdetermined, it may be dangerous to use expert opinion—as the National Academy of Sciences did for Yucca Mountain—as the basis for choosing the best model. In such cases, the best model may be the one determined not only by expert scientific opinion but also by democratic rights. Applied to waste-repository issues, the relevant question is not merely "How safe is safe enough?" but "How safe is equitable enough?" "How safe is voluntary enough?" and "How safe is compensated enough?" In other words, the relevant questions are whether the costs of nuclear waste are borne equitably among all the population, whether communities near radioactive waste facilities voluntarily accept the facilities, and whether they will be compensated for any damages from them.

Such questions suggest that there is an "ethical rationality" appropriate to applied science having potential consequences for human welfare, in addition to the "epistemic rationality" characteristic of model choice in purer science. Scientific rationality, characteristic of pure science, is primarily a rationality of belief that assesses the probability associated with competing hypotheses and the epistemic consequences following from the acceptance of alternative hypotheses. Ethical rationality, characteristic of applied science, is primarily a rationality of action that assesses the ethical goodness or badness (utility) associated with acceptance of alternative hypotheses. For example, suppose one were evaluating a null (no-effect) hypothesis about groundwater migration, an hypothesis suggesting the absence of potential catastrophic effects of nuclear-waste disposal. In such a situation of scientific uncertainty with policy-relevant consequences, scientific rationality might dictate minimizing false positives, whereas ethical rationality might dictate minimizing false negatives (Shrader-Frechette and McCoy 1993, pp. 149–197, esp. 191–197).

2.7 OBJECTIONS

In response to this brief account of model choice in applied science, someone might object that no reasonable scientist or philosopher of science would argue either for using purely scientific rationality or for merely employing the best theory in modeling situations having important policy consequences. However, in explaining rational risk behavior, Laudan (1994, pp. 18–23) repeatedly appeals to the allegedly best theory, that rational risk aversion is proportional to the probability of fatality. Laudan presupposes that risk magnitude is the only relevant norm for risk comparisons and thus ignores issues of consent, compensation, and equity of risk distribution. In using purely scientific rationality in policy-relevant situations, Laudan criticizes the EPA's warnings against passive smoke on the grounds that the risk is small. More generally, Laudan criticizes the "hysteria" and the "hypochondria" of those who are concerned with risks such as dioxin and breast cancer (1994, pp. 5, 23). His second risk "rule" is that each person should determine one's own individual level of acceptable risk and "should not be phobic about engaging in activities with a smaller risk than that one" (Laudan, 1994, p. 167).

In relying on scientific rationality, and in ignoring ethical rationality, even in discussing scientific questions with welfare consequences, Laudan follows the same line of reasoning as many scientists who reduce questions of ethics to questions of science. He assumes that risk probability, and not whether the risk is voluntary or equitably distributed, is all that is needed to determine risk acceptability. G. E. Moore argued that those who attempt to reduce ethical issues (of risk acceptability, for example) to purely scientific issues commit the "naturalistic fallacy". (The naturalistic fallacy is the attempt to reduce ethical questions to purely scientific questions.) Apart from whether Moore was correct—something not at issue here—there is a pervasive and lamentable trend, both in the biological (see Shrader-Frechette and McCoy, 1993, pp. 149–197) and the physical sciences (see Shrader-Frechette, 1993, pp. 99–100), to use purely scientific rationality in situations having ethical consequences, and to extend criteria from one domain (pure science) to another (applied science) for which they may not be well suited.

Another response to this account of model choice in hydrogeology/waste siting might be that it threatens notions of objectivity. Obviously, scientific claims are objective in a way that ethical claims are not. Science, but not ethics, often is empirically confirmable. Arguing for using ethical rationality, as well as scientific rationality, in policy-relevant areas like hydrogeology, means that one cannot tie objectivity to empirical confirmability. Instead, one might tie objectivity to avoiding bias values, to giving even-handed representation of the situation, and to being open to debate.

Scheffler (1982, p. 369) recognized that "objectivity requires simply the possibility of intelligible debate over the merits of rival paradigms." Objectivity requires neither algorithms nor empirical confirmation for evaluation. Because one can be blamed for failure to be objective—in the sense of not being even-handed, of being biased, or of making judgments not open to intelligible debate—it must be possible to be more or less objective in applying ethical rationality.

For example, scientists could be blamed for a failure to be objective if they drew a conclusion about incidence of a disease in the general population yet employed only traditional epidemiological data about white, male subjects. Because people are blame-worthy for not being objective, it is clear either that objectivity (in the sense of avoiding obvious bias) must be attainable or that one can be more or less objective in avoiding bias values.

How might geologists help to guarantee objectivity in the sense of avoiding bias values? Perhaps the best way is to check the predictive or explanatory power of claims and to subject them to review by relevant members of the epistemic community. In other words, they could follow the norms of metascience rather than the norms of naive-positivist epistemology. Although objectivity is not mere consensus, perhaps the best way to recognize objectivity is to use metascience, to subject claims to criticism and analysis by the representative community of knowers.

2.8 CONCLUSIONS

Apart from whether one agrees with this account of scientific and ethical rationality in applications of hydrogeology, it is important to distinguish scientific from ethical rationality. Failure to distinguish ethical from epistemic rationality may mean not only that human welfare could be jeopardized through policy-related scientific decision making, but also that science itself could be hurt. Whenever "best theory" choices, among seriously flawed theories, lead to expert guessing (or decision making by "hired guns" with vested interests) and to confusing scientists' opinions with scientific conclusions, then anti-scientific and anti-intellectual sentiment could be the result. Who wouldn't be anti-scientific when a National Academy of Sciences' committee affirms the reliability of a million-year performance assessment of geological repositories? The anti-scientific, anti-intellectual trend of recent years may be a reaction to scientists who blindly use purely epistemic rationality to select the best theory in a policy-relevant situation—like Yucca Mountain—where there are no good theories. And in policy-relevant areas of science where there are no really good theories or models, choosing default assumptions, rules for decision making under scientific or probabilistic uncertainty, may be just as important as developing science itself. But choosing default assumptions and rules for decision making under uncertainty are, in part, policy decisions. And if they are policy decisions, the public must help make them. No risks without representation.

REFERENCES

Bedinger, M. S. and Stevens, P. R. (eds.). (1990). *Safe Disposal of Radionuclides in Low-Level Radioactive-Waste Repository Sites*. USGS Circular 1036. Big Bear Lake, CA: U.S. Geological Survey.

Blanchard, R. L., Montgomery, D. M., and Kolde, H. E. (1977). *Radiological Measurements at the Maxey Flats Radioactive Waste Burial Site.* EPA-520/5-76-020. Cincinnati, OH: U.S. Environmental Protection Agency.

Carey, W. P., Lyverse, M. A., and Hupp, C. R. (1990). *Hillslope Erosion at the Maxey Flats Radioactive Waste Disposal Site.* WRI Report 89-4199. Louisville, KY: U.S. Geological Survey.

Clark, D. T. and Wilson, B. M. (October 21, 1974). Radiological Health Program, "Project Report, Kentucky Radioactive Waste Disposal Site." Memorandum to Stanley Hammons, M.D. Frankfort, KY: Bureau for Health Services, State of Kentucky.

Eschwege, H. (1976). U.S. General Accounting Office, "Testimony before Congress," in U.S. Congress.

Fleming, G. (1975). *Computer Simulation Techniques in Hydrology.* New York: Elsevier.

Haney, D. C. (1994). "Maxey Flats, Kentucky, Low-Level Nuclear Waste Repository: Past, Present, and Future." *Geological Society of America, Abstracts with Programs 26 (7)*:105–106.

Hartnett, C. (1994). "The Cleanup of Releases of Radioactive Materials from Commercial Low-Level Radioactive Waste Disposal Sites." *Natural Resources Journal 34 (2)*: 349–377.

Hopkins, H., USGS. (1962). "Ground-Water Conditions at Maxey Flats, Fleming County, Kentucky." Unpublished USGS report. Louisville, KY: Department of the Interior.

Hyland, P. et al. (1980). Legislative Research Commission, *Report of the 1978–1979 Interim Special Advisory Committee on Nuclear Waste Disposal.* Research Report No. 167. Frankfort, KY: Legislative Research Commission.

———. (1982). *Legislative Research Commission, Report of the Special Advisory Committee on Nuclear Issues.* Research Report No. 192, Frankfort, KY: Legislative Research Commission.

Laudan, L. (1977). *Progress and Its Problems.* Berkeley: University of California Press.

———. (1994). *The Book of Risks.* New York: Wiley.

Linsley, R. K., Kohler, M. A., and Paulhus, J. L. (1975). *Hydrology for Engineers.* New York: McGraw-Hill.

Lyverse, M. A. (1990). "Results of Recent Geohydrologic Studies at the Maxey Flats Low-Level Radioactive Waste Disposal Site, Fleming County, Kentucky." In M. S. Bedinger and P. R. Stevens, (eds.), *Safe Disposal of Radionuclides in Low-Level Radioactive-Waste Repository Sites.* USGS Circular 1036. Big Bear Lake, CA: U.S. Geological Survey.

Mayo, D. (1997). "Sociological versus Metascientific Views of Technological Risk Assessment." In K.S. Shrader-Frechette and L. Westra (eds.), *Technology and Values.* Lanham, MD: Rowman and Littlefield, pp. 217–250.

Meyer, G. (February 19, 1975). "Maxey Flats Radioactive Waste Burial Site: Status Report" (unpublished report). Washington, DC: Advanced Science and Technology Branch, U.S. Environmental Protection Agency.

———. (1976). *Preliminary Data on the Occurrence of Transuranium Nuclides in the Environment at the Radioactive Waste Burial Site, Maxey Flats, Kentucky.* EPA-520/3-75-021. Washington, DC: U.S. Environmental Protection Agency.

National Research Council (NRC). (1995). *Technical Bases for Yucca Mountain Standards.* Washington, DC: National Academy Press.

Neel, J. (1976). "Statement" in U.S. Congress.

Oreskes, N., Shrader-Frechette, K. S., and Belitz, K. (1994). "Verification, Validation, and Confirmation of Numerical Models in the Earth Sciences." *Science 263*: 641–646.

Papadopulos, S. S. and Winograd, I. (1974). *Storage of Low-Level Radioactive Wastes in the Ground: Hydrogeologic and Hydrochemical Factors.* EPA-520/3-74-009. Washington, DC: U.S. Environmental Protection Agency.

Powell, N. (1991). "A Concerned Community: Plutonium Had Migrated Hundreds of Feet." *Environmental Protection Agency Journal 17(13)*: 31–32.

Sagoff, M. (1985). *Risk Benefit Analysis in Decisions Concerning Public Safety and Health.* Dubuque, IA: Kendall/Hunt.

Scheffler, I. (1982). *Science and Subjectivity.* Indianapolis, IN: Hackett.

Schoeller, H. (1974). "Analytical and Investigational Techniques for Fissured and Fractured Rocks." In R. H. Brown, A. A. Konoplyantsev, J. Ineson, and V. S. Kovalevsky (eds.), *Ground-Water Studies.* New York: United Nations Educational, Scientific, and Cultural Organization, pp. 14.1–14.3.

Shrader-Frechette, K. S. (1993). *Burying Uncertainty: Risk and the Case Against Geological Disposal of Nuclear Waste.* Berkeley: University of California Press.

Shrader-Frechette, K. S. and McCoy, E. D. (1993). *Method in Ecology.* Cambridge: Cambridge University Press.

Todd, D. K. (1980). *Groundwater Hydrology.* New York: Wiley.

Tolkien, J. R. R. (1965). *The Fellowship of the Ring.* New York: Ballantine Books.

Trask, N. J. and Stevens, P. R. (1991). *U.S. Geological Survey Research in Radioactive Waste Disposal.* WRI Report 91-4084. Reston, VA: U.S. Geological Survey.

Travis, John. (December 12, 1998). "Meet the Superbug: Radiation-Resistant Bacteria May Clean Up the Nation's Worst Waste Sites." *Science News 154*, 376–378.

U.S. Congress. (1976). *Low-Level Radioactive Waste Disposal.* Hearings Before a Subcommittee of the Committee on Government Operations, House of Representatives, 94th Congress, Second Session, February 23, March 12, April 6, 1976. Washington, DC: U.S. Government Printing Office.

Walker, I. (September 12, 1962). *Geological and Hydrologic Evaluation of a Proposed Site for Burial of Solid Radioactive Wastes Northwest of Morehead, Fleming County, Kentucky* (unpublished report). Kearny, NJ: U.S. Geological Survey.

Weiss, A. and Columbo, P. (1980). *Evaluation of Isotope Migration-Land Burial.* NUREG/CR-1289 BNL-NUREG-51143. Washington, DC: U.S. Nuclear Regulatory Commission.

Werner, E. (1980). *Joint Intensity Survey in the Morehead, Kentucky, Area.* Louisville, KY: U.S. Geological Survey.

Wilson, K. S. and Lyons, B. E. (1991). *Groundwater Levels and Tritium Concentrations at the Maxey Flats…Site.* WRI Report 90-4189. Louisville, KY: U.S. Geological Survey.

Younker, J. L. et al. (1992). *Report of the Peer Review Panel on the Early Site Suitability Evaluation of the Potential Repository Site at Yucca Mountain, Nevada.* SAIC-91/8001 Washington, DC: U.S. Department of Energy.

Zehner, H. H. (1981). *Hydrogeologic Investigation of the Maxey Flats Radioactive Waste Burial Site, Fleming County, Kentucky.* USGS Open File Report. Louisville, KY: U.S. Department of the Interior, U.S. Geological Survey.

3

Inhabitation and Orientation:
Science Beyond Disenchantment

Bruce V. Foltz

Bruce V. Foltz is Professor of Philosophy and Director of the Senior Honors Program at Eckerd College in St. Petersburg Florida. He also teaches in the Environmental Studies program at Eckerd, and has been a visiting faculty member at St. John's College in Santa Fe, New Mexico. Born and raised in Kansas, where his family had a farm on the Ninnescah River, Foltz received his Ph.D. in philosophy from Penn State University. He is author of Inhabiting the Earth: Heidegger, Environmental Ethics, and the Metaphysics of Nature *(Atlantic Highlands, NJ: Humanities Press International, 1995), as well as a number of articles on environmental philosophy. Foltz is also president, and co-founder, of the International Association for Environmental Philosophy.*

Foltz is concerned here with the ways that scientific knowledge relates to our everyday experience of nature. A long-standing criticism of scientific knowledge is that it provides us with an understanding of things that simultaneously drains the world of its meaning. Science, it is said, reveals a world of objectified facts devoid of purpose and lacking in enchantment. As Wordsworth put it, we murder to dissect.

But must this be true? In contrast to standard accounts, Foltz believes that scientific reason has interests and goals intrinsic to the process of scientific reasoning itself. Most prominent among these is its interest in making us at home in the world. Foltz claims that the goals of learning how to inhabit and orient oneself in the world lie at the roots of the scientific project. Finally, Foltz suggests that the Earth sciences are the scientific disciplines most rooted in our everyday experience of the natural world.

During the darkening years of the mid-1930s in Germany, the philosopher Edmund Husserl worked with a growing sense of urgency on a book-manuscript that he did not live to see published, yet which eventually established him as one of the century's most influential interpreters of modern science. Husserl believed that the modern natural sciences, the pride of European civilization, had arrived at a state of crisis that nevertheless remained strangely unnoticed. He argued that the sciences, originally intending to elucidate the surrounding world of nature and make it intelligible, had unknowingly, but increasingly, become detached from our everyday experience of nature, quietly abandoning it to the dusk of unintelligibility. What had seemed, in the early days of modern science, to be simply a rigorous application of mathematics to the ever-present natural world around us, had gradually—starting with Galileo—begun to substitute for this immediately encountered world of surfaces and smells and colors, of rivers and rocks and vegetation, another world altogether, an idealized, mathematical world that could never be directly experienced at all.

Modern science, Husserl argued, has achieved its successes at the expense of unwittingly performing a kind of subtle shell-game or bait-and-switch, in which the natural world that we encounter everyday—the nature that greets us in an unassuming walk through the woods—has been quietly replaced by a substitute, a mathematicized nature. Worse yet, there has been virtually no attempt to bridge the gap between these two realms of nature—that of mathematicized entities and that of everyday experience—because there is no sustained awareness that the switch has even been made. If Husserl was right about this, there should be no surprise in the suspicion and incomprehension that the sciences experience among the general public, who remain unreflectively encircled by the world of everyday experience, which Husserl termed the *Lebenswelt* or "life-world." The natural sciences, despite and even because of their spectacular results, have left the natural world, which all of us, including scientists, find always about us, as *terra incognita*, uncharted territory whose contours and textures are now explored only by poets, artists, and other culturally marginal figures who are not taken very seriously anyway, precisely because they are not addressing the mathematicized nature of the sciences. And even when the natural life-world is encountered in practical activity—in engineering, in mining, in what is curiously called "development"—the relation to nature is not more direct but less so, as the mathematicized nature of science is further mediated through the mathematical matrix of economics.

Fifty years before Husserl's reflections on the crisis of the sciences, the American writer Samuel Clemens (Mark Twain) reflected on a different, yet strangely parallel crisis in the knowledge of nature. In his autobiographical book, *Life on the Mississippi*, Clemens reflected not on natural science but on his own, personal knowledge of nature gained as a riverboat pilot, and on where it had led him. In contrast to his passengers, who are naively charmed by the beauty of the river, he realizes that his years of knowledge and practical experience now allow him to read it like a book. A gentle dimple on the river's surface is a lovely play of water and light to the superficial eye of the passenger. "It is the faintest and simplest expression the water ever makes," he noted, but "the most hideous to the pilot's eye...for it mean[s] that a wreck or a rock [is] buried [underneath] that could tear the life out of the strongest vessel that ever floated." Having mastered the lexicon of the river, learned how to read the river's surface for signs of what lies beneath, learned to read even the crimson sunset for the bad weather it might bring, he realizes that he has at the same time "lost something which could never be restored to me while I lived. All the grace, the beauty, the poetry had gone out of the majestic river...[A] day came when I [stopped noticing] the glories and charms which the moon and the sun and the twilight wrought upon the river's face.... The romance and beauty were all gone from the river. All the value any feature had for me now was the amount of usefulness it could furnish toward compassing the safe piloting of a steamboat." Woefully, he extrapolates his conclusion about the cost of such knowledge: "Since those days, I have pitied doctors from my heart. What does the lovely flush in a beauty's cheek mean to a doctor but a "break" that ripples above some deadly disease? Are not all her visible charms sown thick with what are to him the signs and symbols of hidden decay? Does he ever see her beauty at all...? And doesn't he sometimes wonder whether he has gained most or lost most by learning his trade?" (Clemens, 1990, p. 265).

Within a decade of Mark Twain's reflections on knowledge and loss, two of his American contemporaries addressed the same problem, but with respect to the natural sciences of their time. In a journal entry of May 14, 1888, published in his prose work *Specimen Days*, the poet Walt Whitman wrote that although he loves to listen to the bird songs around his home in the New Jersey woods, he does not know the names of many of the birds he hears, nor does he even seek to know such things. "You must not know too much, or be too scientific about birds and trees and flowers.... A certain free margin and even vagueness—perhaps ignorance, credulity—helps your enjoyment of these things, and of the sentiment of feathered, wooded, river, or marine Nature generally. I repeat it—[I] don't want to know too exactly, or the reasons why" (Whitman, 1990, p. 246).

A few years later, the great naturalist John Burroughs wrote, more temperately and less stubbornly than Whitman: "I am not always in sympathy with nature-study as pursued in the schools, as if this kingdom could be carried by assault. Such study is too cold, too special, too mechanical; it is likely to rub the bloom off Nature. It lacks soul and emotions; it misses the accessories of the open air and its exhilarations, the sky, the clouds, the landscape, and the currents of life that pulse everywhere.... When we look upon Nature with fondness and appreciation she meets us halfway and takes a deeper hold upon us than when studiously conned. Hence I say the way of knowledge of nature is the way of love and enjoyment, and is more surely found in the open air than in the schoolroom or the laboratory." (Burroughs notes that his own knowledge of nature—by today's standards, an impressive grasp of what would now be called ecology—had come not through studying nature "with notebook and field glass in hand," but effortlessly, unintentionally, as he was absorbed in nature "while fishing or camping or idling about.") He concludes not with the willful ignorance that Whitman embraces, but with this exhortation: "to enjoy understandingly, that I fancy, is the great thing to be desired" (Burroughs, 1990, p. 275).

Is the scientific knowledge of nature inherently destructive of our appreciation of the beauty and goodness of the natural world around us? Does it necessarily lead us away from the nature that can charm and delight us, that can make us feel grateful to be inhabitants of the Earth? An additional text, from a contemporary scientist, will be helpful in focusing these questions. Lewis Thomas, biologist and immunologist, and long-time president and chancellor of New York's Memorial Sloane-Kettering Cancer Institute, wrote of his experience at Tucson's Desert Museum watching, at eye level, the play of otters and beavers in glass tanks. "Transfixed" by the play of these splendid animals, filled only with what he calls "pure elation mixed with amazement at such perfection," he becomes apprehensive about what will happen as soon as he starts to analyze scientifically what he is seeing: "I wanted no part of the science of beavers and otters; I wanted never to know how they performed their marvels; I wished for no news about the physiology of their breathing, the coordination of their muscles, their vision, their endocrine systems, their digestive tracts. I hoped never to have to think of them as collections of cells. All I asked for was the full hairy complexity, then in front of my eyes, of whole, intact beavers and otters in motion." But of course, as soon as he begins to feel this, the spell is broken, not by thinking about the biology of otters, but by analyzing what part of his own brain is producing this very experience, perhaps the brain stem, perhaps his limbic system at work. At that moment, he mourns, "I became a behavioral scientist, an experimental psychologist, an ethologist, and in the instant I lost all the wonder and the sense of being overwhelmed. I was flattened," he concludes, "back in the late twentieth century, reductionist as ever, wondering about the details by force of habit" (Thomas, 1990, p. 584).

In these accounts, scientific knowledge is not the perfection of wonder, the fulfillment of our delight in the details of the earthly, but their destroyer, the enemy of beauty, the kidnapper and ravager of lived, embodied, nature, disillusioning us of the charms and enchantments we had once cherished. Nor can we comfortably dismiss them, for these personal and anecdotal accounts intersect too closely with the analysis of Max Weber, contemporary of Husserl and immensely influential as a sociologist of science. Weber argued that Western rationality, and especially modern science as its finest exemplar, has resulted in a uniquely modern relation to the world around us, which he called (borrowing the term from Schiller) the *Entzauberung*, the "disenchantment" or literally the "de-magification" of the world. "The fate of our times," he argued, "is characterized by rationalization and intellectualization and, above all, by the 'disenchantment' of the world" (Weber, 1958b, p. 155). Science has stripped away the magic of the world, by showing that nature is everywhere subject to technical mastery and control. The disenchantment of the world, Weber wrote, does not mean that we already understand everything about the world; it means only that we have become convinced that there are, in principle, "no mysterious, incalculable forces that come into play, but rather that one can [at least] in principle, master all things by calculation."

This means first of all that we no longer need have recourse to "magical means in order to master or implore the spirits" (Weber, 1958b, p. 139). But beyond this, it also implies that the world is bereft of its own meaning and value, making it questionable what the point would be in mastering it at all. "Wherever rational, empirical knowledge has consistently worked through to the disenchantment of the world and its transformation into a causal mechanism," Weber claimed, it encounters and refutes the claims of "every intellectual approach which in any way asks for a 'meaning' of inner-worldly occurrences" (Weber, 1958a, p. 350). Weber was aware of the harshness of his conclusion, that scientific "progress" has resulted in rendering pointless that very progress itself in mastering a disenchanted and insignificant world, and he was fully aware that these views also imply that science itself does not possess "any meanings which go beyond the purely practical and technical." But he nevertheless counseled courage and intellectual honesty in continuing the pursuit of science to those who can withstand this disillusionment, while to weaker and less intrepid souls he sardonically recommended theology (Weber, 1958b, pp. 139, 150).

Can we avoid these unhappy conclusions? Must we contrive, with Lewis Thomas, to suppress our knowledge of nature, if we are to enjoy a day at the zoo? Should we, with Whitman, avoid science altogether in the first place? Or is there hope in Husserl's call to build links between the world of scientific concepts and the life-world of nature around us? Is it possible, in Burrough's words, to enjoy nature understandingly, that is, to preserve and even enhance the charm and beauty and enchantment of nature by in some way combining it with scientific understanding? For what indeed is the point of mastering, or even understanding, a world that is drained of its meaning? A closer examination of the connection between scientific understanding and the technical mastery of nature can help clarify what is at issue here.

The relationship between science and mastery has been explored at length by the German philosopher and sociologist Jürgen Habermas in his work, *Knowledge and Human Interest* (1972), and in which Habermas borrows from Kant the notion that reason has its own, native "interests." According to this view, rationality in general has interests that are intrinsic to its legitimate operation—interests (or, we might say, "concerns") that are proper and necessary to reason itself, rather than being exterior or ulterior forces. When it is argued, to the contrary, that reason should be disinterested, this should be understood only to mean that reason must resist powers extrinsic to it, such as the passions that might serve to warp or sway its proper functioning.

Habermas, then, departs dramatically from the view that has been customary in Western epistemology since the ancient Greeks, and which sees knowledge as theory, as a pure, detached "seeing" or onlooking. (Indeed, the Greek word *theôria* is related to the English word "theater," and originally evoked the activity of a viewer or spectator.) Habermas is, rather, closer to the American pragmatist philosophers, understanding reason to be the means that humanity employs to reproduce the conditions of its existence and to build up a cultural world within which it can be at home. Accordingly, Habermas distinguishes three kinds of rationality, each possessing its own characteristic interest or concern. First, there is a technical, instrumental, or work interest in mastering the natural world, in order to produce the means of human sustenance. This kind of rationality is employed in the *natural sciences*. Second, there is an interest in human interaction within the cultural world that is shaped by language. Its corresponding mode of rationality is characteristic of what American universities call the *humanities*. Finally, there is a third human interest in freedom or emancipation, whose task includes the critical examination of the other two modes of knowledge, to ensure that their claims to knowledge are not in fact disguised and distorted claims to power over other human beings. Habermas finds this last interest embodied in the *social sciences*, and thus sees them as having a very different interest than the natural sciences, an interest not in mastery but in emancipation.

What, then, are the implications of Habermas's analysis for our understanding of the natural sciences and the problem of disenchantment? In one sense, science is bound even more tightly to the mastery of nature here than with Weber. Indeed, Habermas sees the positivist or objectivist understanding of natural science—that it is simply a disinterested mirror of reality, an accurate picture of a preconstituted world of "facts"—as one of the prime targets of critical or emancipatory rationality, for it obfuscates the crucial questions of how, and for whom, science is serving the shared human activity of work. Yet through being subordinated to the goal of human emancipation, the natural sciences gain a restoration of meaning. In contrast to Weber, science for Habermas properly pursues not mastery for its own sake, but mastery for the sake of serving human needs, as established through a process of rational and undistorted communication. Science receives a meaning here, but it is one that puts us back on the river with Samuel Clemens, whose efforts to secure the safety of his passengers left him with a disenchanted river, whose only meaning lay in the work it was his duty to perform.

When we turn to the issue of science's role in the disenchantment of nature, there are several questions to be asked concerning Habermas's idea that the interest of the natural sciences consists solely in the technical mastery of nature. First, should nature receive its significance only from the sphere of human needs and desires? Is nature essentially meaningless in its own right? Second, is there nothing in science corresponding to the ancient concept of *theôria*, a pure beholding to be undertaken for its own sake? Is it merely atavistic to speak of a contemplative interest in nature? Third, is the sphere within which human beings are properly at home exclusively a cultural sphere, a realm of language and texts and symbols? Are we ever truly at home if we are not also at home in the natural environment, at home not just in the sense of having mastered it, but in being able to find hospitality amid the beauty, the beneficence, and the wild wisdom of nature? Finally, is the notion of a work interest truly primary? Is it self-explanatory? Or is the notion of work for its own sake, even when it is understood as serving human needs and desires, no more coherent than that of mastery for its own sake?

What is the interest of scientific reason? This is not a psychological or biographical question, but rather a logical and epistemological one, a question concerning the rationality that is deployed in science. Thomas Kuhn, author of *The Structure of Scientific Revolutions* (1971), rejects the notion of a work interest, arguing instead that scientific inquiry is animated by a basic interest in solving puzzles, and that many important real-world problems are skipped over by science precisely because they do not possess the most important feature of any good puzzle—that is, they must admit of a solution. Scientific revolutions occur, he maintains, when the problem-solving activity becomes frustrated, and it is found that only a new paradigm of what is to count as science can allow these puzzles to be solved successfully. This would seem to turn scientific inquiry into a strange and elaborate game, in which nothing is ever really gained because there is no real partner on the other side.

What is it that animates a scientific revolution? If we look at the writings of those figures who have played key roles in radically changing the nature of science—of Galileo and Copernicus and Newton, of Darwin and Hutton and Einstein and Heisenberg—do we not find instead a sense of wonder and amazement at the dawning of a new sense of what the world is like, an excitement or perhaps even dread at the prospect that the world may not be the kind of place we had thought it to be? This sense of wonder is neither an eagerness to undertake new forms of work and mastery, nor a delight in discovering new puzzles, but instead an astonishment at finding oneself inhabiting a new kind of place, at finding out that one is not residing in quite the same kind of universe that it had seemed to be, in discovering that one is living in a new world. Is it, after all, an accident that Europeans were discovering and exploring what to them were new worlds of territory at the same time that scientists began to discover that nature itself had become a new kind of world to them?

Novelists and screenwriters have often portrayed this sense of a new world in their depiction of amnesiacs regaining consciousness, and of time- or space-travelers finding themselves in new worlds. What are the questions that such fictional figures ask first? They ask: "Where am I?" "What kind of place is this?" and "What time is it?" But we don't have to be time-travelers, or initiate scientific revolutions, in order to pose these questions. Let us simply imagine a camper, hiking through a wild area, perhaps along the banks of Samuel Clemens's Mississippi, if there is any wilderness left along its banks, strolling and idling as Burroughs commends, just to enjoy the country. What kind of cognitive activity is going on? Isn't it a continual orienteering, an asking of these same questions: "Where am I?" "What kind of place is this?" and "What time is it?" And when it is time to make camp, the orienteering becomes more systematic and focused, becomes a *reconnaissance*, which nevertheless asks more earnestly those same questions. There is, of course, work to be done in setting up camp, a tent to be put up, a fire to be built, and so on. But both logically and epistemologically, the reconnaissance is more primary and has to come first. Is this a good place to stay? Is it too early to make camp? Is water close by? Is there danger of flood? Will it be too sunny here in the morning? Will the blackberry bushes attract bears? And so on. The work makes no sense, cannot even be defined, until the reconnaissance is at least provisionally complete, until we have gotten our bearings, for only then will it be clear what has to be done.

The camping trip offers us a good analogy for our human sojourn upon Earth. Is it not generally the case that the work we have to do, as well as the requisite puzzle-solving, are always derivative from the reconnaissance? *Where are we?* Where are we in this vast universe? Where are we in the strange tapestry of evolution? Where are we in the shifting trajectories of plate tectonics? *What kind of place is this?* Are humans inevitably at odds with their environment, or is a more friendly relationship the more salutary course? Are we sitting ducks for hostile microorganisms of all kinds, or perhaps a random meteor, or can we usually count on the resources of our own immune systems and the laws of probability? What kind of place is this after all? And finally, *what time is it?* How long have we humans been around? How long has this biome, this mountain, this climate, this Earth been here? How long does this old planet of ours have left? How are we situated not just spatially, but temporally within the macro-changes of Earth and sky?

This is the kind of reconnaissance that all human beings, whether of the Neolithic or of the third millennium, must continually be making, for our most essential interest lies in *knowing what kind of world we inhabit*. These are not abstract, second- or third-order questions we ask if we have the leisure, but the most important and most basic questions, the ones whose answers (however provisional) define the work there is to be done, by showing us what kind of world we live in. These questions of reconnaissance are not only not impractical, but they lay the foundations for the practical itself, because they establish the parameters of all our needs and tasks.

Beyond this, they can also open up new strata of practical questions and commitments. For example, if science serves primarily an inhabitory interest of human reason, then that interest is fulfilled not just by reconnoitering, but by making sure that there is good country to survey, country that is good to inhabit. That is, just as medical science must understand itself as pursuing not just a disinterested knowledge of the human body, but a knowledge that promotes human health, so too the natural sciences in general could see their goal not just as disinterested knowledge of nature, but as knowledge that serves to promote a healthy and habitable environment. What research needs to be done, what knowledge needs to be acquired, would thus be ordained by what knowledge is needed to promote a healthy, and a beautiful, natural environment, as well as by what knowledge we need in order to stay oriented within that environment. Put differently, if we think through to its conclusions the idea of science as reconnaissance, we can see that it has an ultimately *ethical* bearing—not in the narrower sense of being concerned with rights and duties, but in the broader sense that its project of orienting us must at the same time be attuned to the well-being of its surroundings. Science must thereby concern itself with the normative aspects of knowledge, for health and integrity and beauty are modalities of goodness, and actions that promote the good are by definition ethical. Nor should this be surprising if we consider that the word "ethics" comes originally from the Greek *ethôs*, which meant *lair, dwelling-place, place of inhabitation*. Concern for the well-being of a place is one of the things that makes us inhabitants of that place, as opposed to mere passersby.

At the same time as they open up the practical-ethical realm, the questions of scientific reconnaissance are profoundly contemplative as well, questions to be pursued for their own sake, to be pondered over and wondered about. And because they are contemplative, they connect directly to those same questions as they are pursued in philosophy, in art, in literature, in history, and in theology, each of which undertakes, in its own way, a similar project of reconnaissance. Moreover, because they address themselves primarily to the lived interface between humanity and the world, they are questions that connect all the details of scientific inquiry back to the life-world we encounter everyday, linking science back to the inhabited world from which it always proceeds.

The questions of reconnaissance—questions of orientation in the interest of inhabitation—are the basic questions that animate all our knowledge of the physical world, and as such they are the basic questions of science as well. We pose these questions because we are inhabitants of Earth, dwellers upon Earth; it is our place of residence, and we need and want to know what kind of place it is, not only to find out what work there is to do, but also simply because we want to know the neighborhood in which we reside—we want to be at home. In its primary relation to the physical world around us, reason possesses an inherent interest in inhabitation, an elemental *inhabitory interest*, to which an unceasing reconnaissance is called to minister.

If this is the case, then Husserl's project of connecting the nature of mathematical science with that of the life-world would lend itself to a variegated and ongoing fulfillment. There is no facet of nature that could be safely ignored by a reconnaissance which would make Earth truly inhabitable, for there is no feature or nature so insignificant that it might not potentially alter our understanding of the whole, and there are no findings of science without potential bearing upon such an endeavor. If the most basic task of the natural sciences lies in this kind of reconnaissance of the universe, then it has a further task of showing in each case not only what kinds of technical mastery its results imply, but more importantly how these findings help us get our bearings in the world in which we all reside. Finally, this task must not be seen as extra-scientific, as mere "popularization," but rather as the most important stage of all in the process of scientific inquiry, as the final, indispensable step that actually scores the points.

If science has at its heart this inhabitory interest, then the magic and enchantment of the world need not be lost at all, for it is continually evoked and even deepened in the very asking of those questions: "Where are we?" "What kind of place is this?" and "What time is it?" Nor should it be forgotten that these are the same questions that the philosopher and poet ask as well. And if science shares this same inhabitory interest with the philosopher and poet, the theologian and historian, then it is with these that the scientist shares the closer kinship, rather than with the engineer, the miner, and the developer, whose pursuits are directed toward the derivative interest of work. Indeed, until the last century, no one questioned this kinship of science with the arts and humanities. It is high time that it be restored, for it is nothing less than Earth itself that is at stake, and it is nothing less than a planet that is *simultaneously* meaningful *and* intelligible that is likely to be saved and preserved with whatever intactness remains possible to it. Not only must the kinship be restored, but links of all kinds must be made explicit between the understanding of nature in the sciences and the humanities if our orientation, and inhabitation, is to be integral.

Perhaps too the Earth sciences, whose reconnaissance is always ultimately directed toward the lived interface of humanity and nature—toward mountains and rivers, toward cold fronts and droughts—have a special role to play in that restoration. The biochemist might be able to see the world through a microscope indefinitely, as the physicist can at least possibly subsist in a world of numbers and formulas and models without coming up for air, but the Earth scientist can never afford to lose touch with the Earth at our feet and the sky above us, for it is always just these that the Earth sciences must elucidate.

Finally, if scientific rationality is ultimately based upon an inhabitory interest, then it would find the ultimate satisfaction of that interest in a kind of contemplation, in *theôria* in the truest sense: in residing as an inspired onlooker of the spectacle. There is much evidence of this to be found in the reflective writings of individual scientists, and it is articulated at a second order of reflection in a text by Michael Polanyi, who has written extensively about the practice of science. In his book *Personal Knowledge* (1964), Polanyi argues that when we use scientific theories we reside within them, something that he calls "in-dwelling." He goes on to suggest that "a true understanding of science and mathematics includes the capacity for a contemplative experience of them, and the teaching of the sciences must aim at imparting this capacity to the pupil" (Polanyi, 1964, p. 196). Rather than simply employing the theories in order to gather positive knowledge of nature, we can contemplate the theories themselves much as the religious mystic contemplates the divine. But if it turns out that science issues from an inhabitory interest in the first place, then scientific theory is not just a set of useful ideas, but a survey view of the terrain—not a closed system of concepts, but a bearing and comportment within nature, a dwelling not within concepts but within the world. It was noted earlier that the ancient Greek word from which both "theory" and "theater" derive *(theôr)* means to look upon a spectacle. What Polanyi describes, then, would be better understood as a stepping back from the immediate, and secondary, work interest of science into the more primary inhabitory interest, letting the theories be a theater upon the world we inhabit. Or put differently, in the midst of a never-ending reconnoiter, we need to stand back and enjoy the view that has been gained so far, to get a sense of the country we inhabit. Science's underlying contemplative component was unquestioned until recently; perhaps it is time that this too be reasserted and reclaimed by the natural sciences.

So understood, scientific knowledge is not only useful but also essential for a fully human inhabitation of Earth. Without an understanding of our surroundings, we are literally disoriented—as are too many of us too much of the time in relation to the Earth upon which we reside and the heavens beneath which we pass through the course of our days. Comprehended and articulated as serving this inhabitory interest, science would neither diminish our experience of nature nor disenchant it, but offer us the capacity to fulfill John Burroughs's injunction to enjoy it understandingly.

REFERENCES

John Burroughs, *Time and Change*. In *The Norton Book of Nature Writing*, edited by Robert Finch and John Elder (New York & London: W. W. Norton & Co., 1990).

Samuel Clemens, *Life on the Mississippi*. In *The Norton Book of Nature Writing*, edited by Robert Finch and John Elder (New York & London: W. W. Norton & Co., 1990).

Jürgen Habermas, *Knowledge and Human Interest* (Boston: Beacon Press, 1972).

Edmund Husserl, *The Crisis of European Sciences and Transcendental Phenomenology*, translated by David Carr (Evanston, IL: Northwestern University Press, 1970).

Thomas S. Kuhn, *The Structure of Scientific Revolutions*, Vol. 2, No. 2 of *International Encyclopedia of Unified Science*, edited by Otto Neurath (Chicago: University of Chicago Press, 1971).

Michael Polanyi, *Personal Knowledge: Toward a Post-Critical Philosophy* (New York: Harper & Row, 1964).

Lewis Thomas, *The Medusa and the Snail*. In *The Norton Book of Nature Writing*, edited by Robert Finch and John Elder (New York & London: W. W. Norton & Co., 1990).

Max Weber, "Religious Rejections of the World and Their Directions." In C. Wright Mills and H. H. Gerth, editors, *From Max Weber: Essays in Sociology* (New York: Oxford University Press, 1958a).

Max Weber, "Science as Vocation." In C. Wright Mills and H. H. Gerth, editors, *From Max Weber: Essays in Sociology* (New York: Oxford University Press, 1958b).

Walt Whitman, *Specimen Days and Collected Works*. In *The Norton Book of Nature Writing*, edited by Robert Finch and John Elder (New York & London: W. W. Norton & Co., 1990).

4

The Modern Earth Narrative:
Natural and Human History
of the Earth

Richard S. Williams, Jr.

Richard S. Williams, Jr., is a senior research scientist with the U.S. Geological Survey in Woods Hole, MA. He holds a BS and MS in geology from the University of Michigan, and a Ph.D. in geology from Penn State University. He has worked for the Atlantic Richfield Company, Raytheon, and HRB-Singer, and is also Vice Chairman of the National Geographic Society's Committee for Research and Exploration. One of his many research areas is the relationship between volcanic activity and glacial fluctuations in Iceland.

In this essay, Williams offers a summary account of "the state of the planet" from a scientific point of view. Williams's narrative embraces a classically modernist attitude toward the role of scientific knowledge in resolving societal controversies. This view holds that scientific facts provide an objective standpoint for the formulation of public policy. Scientific information and perspectives are not only necessary, but also sufficient for directing the community. For once the scientific facts are known, rational, self-interested individuals will come to similar if not identical positions on questions like global warming and the loss of biodiversity.

This modernist position—that sound science directly leads to rational public policy—remains the dominant stance within both scientific and public-policy circles. It rejects the belief that all knowledge, scientific and otherwise, is fundamentally interpretive in nature, which would make it possible for reasonable people to come to profoundly different conclusions. It denies as well the claim that scientific research itself is always grounded in various types of methodological value judgments. The modernist position also puts a premium on science education: Along with a call for further scientific research, the modernist's goal is to get scientific facts out to the public so that people can make rational choices based upon scientific facts. This essay thus provides a powerful contrast to Daniel Sarewitz's essay in this volume.

The scientific story of the natural and human history of Earth, a modern Earth Narrative, is a story constructed on the two factual pillars of Deep (or Geologic) Time and biological evolution. This story provides humans with an objective perspective of the 4.5-billion-year history of the Earth system, including the place of humans in it and in the cosmos. In this essay, the natural variability of global environmental changes and human changes in Earth's biosphere and geosphere are presented in the context of the continuing rapid increase in human population and the increasingly negative impact of human activities on Earth's ecosystems. My claim is that these facts provide the basis for sound decision making concerning our and the planet's future.

The twenty-first century will be a pivotal time in the fate of Earth's biosphere. Whereas human modification of the geosphere will slowly recover over time, human changes to the biosphere are a far more serious matter: Extinction of species is forever. Will humans effectively use our new knowledge of Earth's natural and human history to stop further degradation of Earth's ecosystems and extinction of its biota? The fate of Earth's biosphere, including *Homo sapiens*, depends on an affirmative answer to this question. The fate of Earth's biosphere also depends on the affirmation by all humans of the global importance of conservation of the planet's biotic heritage.

4.1 EVOLUTION OF LIFE ON EARTH

EARTH MATTERS

The Earth System consists of Earth's geosphere (the solid Earth), atmosphere, hydrosphere (liquid water), cryosphere (frozen water), biosphere (all forms of life), and its climatic processes, hydrologic cycle, and biogeochemical cycles (Figure 4.1). This system has always mattered a great deal to Earth's diverse and numerous terrestrial and marine life forms. Earth provides a dynamic and changing planetary environment that is home to a great variety of ecosystems. Earth is unique among all the planets, moons, asteroids, and comets of our Solar System in supporting millions of different types of living organisms. Some forms of life may exist or may have existed on Mars or on other planets and moons. Nonetheless, as far as is presently known, Earth is the only body in the Solar System to harbor life. Moreover, Earth has had an environment suitable for life-forms continuously for more than 3 billion years.

Over time, life on Earth has followed an evolutionary, upward trajectory toward greater diversity of species. Life has evolved from a few marine families in the Proterozoic era to 100 families in the Middle Cambrian period and more than 900 families today, while vertebrate brains have increased in complexity. This evolutionary pattern has been interrupted by five major extinction events since multicellular life first appeared in abundance in the early Middle Cambrian about 530 million years ago. These five major extinction events are thought to have been caused by some combination of large meteorite impacts, major climate-change events, and prolonged volcanic eruptions. A sixth major extinction is probably already underway, but this time, unlike past natural mass extinctions, the cause is definitely "terrestrial"—through *human modification* of the geosphere and biosphere.

4.2 HISTORICAL DEVELOPMENT OF THE EARTH NARRATIVE

The scientific story of the natural and human history of Earth, the so-called Earth Narrative, is a compelling one. This essay claims that its widespread acceptance by humans is absolutely mandatory if we are to attain the goal of the conservation of *all*

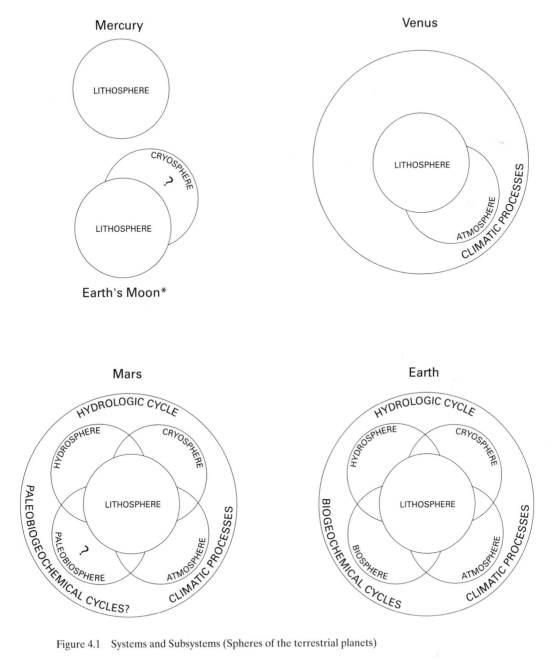

Figure 4.1   Systems and Subsystems (Spheres of the terrestrial planets)

of Earth's living organisms. This goal must be accomplished because of the rapid increase in human populations worldwide and because of the associated environmental changes (local, regional, and global) in the Earth system, especially the alteration and the degradation of its ecosystems.

The scientific narrative of Earth has been painstakingly compiled and carefully formulated by many scientific disciplines during the past two centuries. However, only during the latter part of the twentieth century have the critical missing elements of the story been developed, such as plate tectonics, radioactive dating (geochronology), deoxyribonucleic acid (DNA), and the evolutionary history of hominids. These previously missing elements have revealed a coherent and factual natural and human history of Earth. As Carl Sagan noted in his last book, "The twentieth century will be remembered for three broad innovations: [medical advances, weapons of mass destruction]…and unprecedented insights into the nature of ourselves and the Universe. All three of these developments have been brought forth by science and technology…" (Sagan, 1997, p. 204).

This new Earth Narrative is intellectually far more exciting and logical than the various earlier mythological narratives of Earth's creation. The early myths were first expressed orally by primitive humans as they tried to explain the world about them, especially the many seemingly inexplicable natural events and processes: the changing seasons, lightning, thunder, phases of the moon, comets, eclipses, and stars in the night sky, for example. Many of these mythological and fanciful narratives, which are part of virtually all cultures and religions, were later written down. The narratives that have survived into the modern era have normally been part of a religious text, such as the Old Testament, which is used by the religions of Judaism and Christianity. Or they have been an important cultural element of a group of humans, such as the Aborigines of Australia.

The Aborigines, the first humans to settle Australia and Tasmania, are one of the world's oldest continuous cultures, at least 40,000 years old. In caves and rock shelters, Aboriginal painters have depicted, for at least 20,000 years, various animals and mythological beings from the Dreamtime, the ancient time of Earth's formation. Aborigines believe that humans, animals, and plants are all related to one another (scientifically correct from a DNA viewpoint) and to the environment. A complex story of the creation and development of what we now call the Earth system is embodied in Aboriginal myths.

An example of a complex story of the creation of Earth and the Universe that is not part of a modern-day religion but has survived into modern times is Norse mythology. The Norse creation myth, once part of a pagan religion, was saved by the great Icelandic poet and historian Snorri Sturluson (1179–1241) in *The Prose Edda*. Sturluson wrote this myth down in the thirteenth century, two centuries after Iceland was converted to Christianity. In his account of pseudoscience, Shermer lists six different categories of creation myths embraced by various cultures (Shermer, 1997).

The scientific narrative of Earth did not get started until the latter part of the eighteenth century, when natural historians began to make direct field observations of Earth. The founder of the science of geology was the Scotsman James Hutton. Hutton was among the first scientists to recognize the great length of time that geologic processes must have acted in order to create Earth's rock formations and landscape. He also introduced the concept of "uniformitarianism," which means that processes active today must have acted similarly in the past over very long periods of time. How incredibly long this period of time stretched was not established until the latter half of the twentieth century, when geochronologists, using the known rates of decay in radioactive elements, determined the age of Earth's oldest rocks to be about 3.8 billion years in Greenland and Australia.

In fact, the greatest single contribution of geology to human understanding of the natural history of planet Earth and the universe is the concept of Deep or Geologic Time. When combined with the concept of biological evolution, set forth by the biologist and geologist Charles Darwin in 1859 in *The Origin of Species*, the scientific explanation for the biological component of the natural history of Earth began to take shape. These fundamental concepts are intertwined, because an awareness of Deep Time is a necessary requirement for the long course of biological evolution of life on Earth to take place.

The nonintuitive concepts of Deep Time and biological evolution provided the two pillars upon which twentieth-century scientists wrote the first draft of the natural and human history of Earth. These scientists used sophisticated analytical tools and intensive fieldwork that ranged over both the land and in the oceans. Essentially correct in its presentation of the basic framework of the history of the Earth System, much more scientific knowledge still needs to be discovered to add greater detail to this already complex yet logical framework. Also still to be resolved is the magnitude of the growing impact of a single species, *Homo sapiens*, on the Earth System, and the planetary impact of the activities of 6 billion humans on Earth's geosphere and biosphere, a population increasing at the rate of 90 million each year (Figure 4.2). The paleoanthropologist Richard Leakey and the science journalist Roger Lewin correctly refer to the growing human impact on Earth's biosphere as *The Sixth Extinction* (Leakey and Lewin, 1995).

4.3 HUMAN CHANGES IN THE GEOSPHERE AND BIOSPHERE

Human changes in the geosphere are probably reversible, given sufficient time; however, some human changes to the biosphere are *not* reversible. Every extinction of a species means that its unique assemblage of genes will no longer be part of the biosphere, nor will any new species emerge or radiate from that specific genetic line. Therefore, its unique genetic line is permanently truncated. For example, four birds—the great auk of the North Atlantic Ocean, the heath hen of the northeastern United States, the passenger pigeon of the central United States, and the moa of New Zealand—as well as thousands of other species of animals and plants have become extinct in the past few hundred years. These extinctions have been caused by hunting by humans, introduction of diseases by humans, human alteration of ecosystems, or other modifications of the natural landscape. Birds are particularly vulnerable to human activity: All of the approximately 10,000 known species of birds occupy specialized ecological niches. In addition, migratory birds must contend with increasingly altered ecosystems in their summer breeding habitats, in their winter homes, and in their transient stopover habitats, where they rest and refuel.

The state bird of Hawaii, the Hawaiian goose, or nene, which evolved from chance arrivals of Canadian geese, has been reduced to about 400 individuals on the

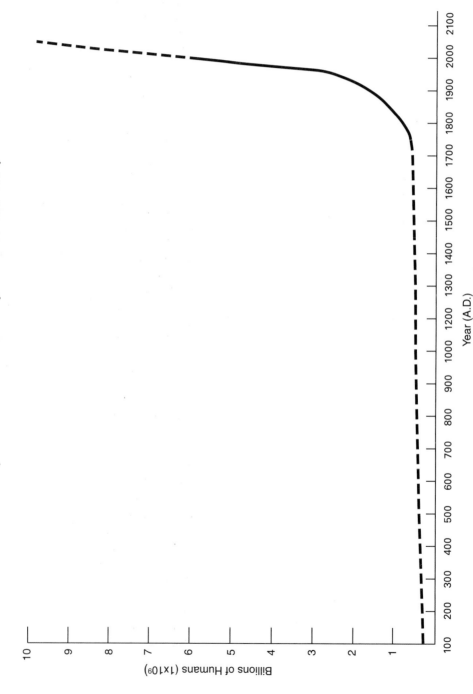

Figure 4.2   Postulated Past, Present, and Projected Human Population of Planet Earth (100–2050 A.D.)

big island of Hawaii and by another 400 or so in and around Haleakala Crater on Maui (the latter group was reintroduced into the wild after captive breeding). The mongoose, an intentionally introduced predatory mammal that is endemic to India, is a constant threat to the survival of eggs and young goslings of the nene (and also to other ground-nesting birds). Of 67 endemic birds in Hawaii that were documented in the late eighteenth century, 25 are probably extinct now, and another 28 are endangered or rare. "Biologists estimate that species are being lost hundreds or even thousands of times faster than normal. Whatever the cause of earlier extinctions, the sixth major extinction is the first mass dying driven by human activity" (Swerdlow, 1999).

4.4 AGRICULTURE AND EXPANSION OF HUMAN POPULATIONS

The last great ice age ended about 10,000 years ago, marking the beginning of the latest interglacial known as the Holocene epoch. It is in the Holocene that agriculture was invented by humans. Until that time, at the dawn of human civilizations, all humans were hunter-gatherers. Earth's total population of hunter-gatherers 10,000 years ago has been estimated by the United Nations Population Bureau to have been approximately 200 million—about the total population of the United States in 1970!

Agriculture was the catalyst that propelled humans into the position of both the dominant species on the planet and the major modifier of the Earth System (Diamond, 1997). The development of agriculture in the Middle East (wheat and other grains) and in Asia (rice) eventually spread to other regions. In the Western Hemisphere agriculture developed independently. As an assured source of food became available, larger families resulted and populations grew. Along with agriculture came extensive modification of the land. This was both intentional (e.g., deforestation, irrigation systems, development of a transportation network for food distribution) and unintentional (e.g., soil erosion, lowering of the water table, and soil salinization). In addition, modern agriculture also uses a great variety of chemicals such as insecticides, herbicides, and fertilizers to maintain or increase production yield. These chemicals ultimately enter surface and groundwaters.

The rapid population increase during the last century is the result of advances in food production, medicine, and public health. These advances include vaccination against infectious diseases, the widespread use of drugs such as antibiotics to combat bacterial infections, and the development of potable water supplies (chlorinated public water-distribution systems). Interspecies transfer of viral diseases to humans (for example, Swine flu and AIDS) poses a continuing challenge to epidemiologists. In fact, emerging infectious diseases may be the ultimate natural limit to the maximum total human population on Earth. Populations of Earth's other animals are limited in nature by available food supply, predators, access to surface water, disease, physiological limits to temperature extremes, migratory behavior, degree of fertility, and social behavior patterns. Humans, through the application of various technologies, have succeeded in overcoming all these natural limitations, while maintaining a very high level of fertility. Other than the emergence of a widespread fatal disease or diseases, a natural occurrence perhaps made worse by dense human populations and unable to be controlled by modern medicine or public-health interventions (e.g., behavioral changes), only a conscious and willful decision to reduce human fertility will limit human populations. However, limitations of our resources (food and water) may also eventually play a role.

What will be the ultimate human population on Earth? Will it reach current United Nations' projections of 11 billion by 2060, or will it be much greater? The present rate of population growth (1.4 percent annually) forecasts a doubling to 12 billion by 2050! A major scientific question concerns the means by which the human population will be limited and at what standard of living or quality of life. Will it be by rational means, or by tragic ones such as those forecast in the late eighteenth century by the Rev. Thomas Malthus: poverty, starvation, war, and/or disease?

4.5 NATURAL VARIABILITY IN THE EARTH SYSTEM

Let us now turn our attention to a summary of the current scientific knowledge of the Earth System from the perspective of the geosphere and biosphere, climatic processes, hydrologic cycle, and biogeochemical cycles. Of particular scientific importance is the ability to distinguish accurately using statistical probability the natural variability in the Earth System from the changes caused by human activities. The global study of the Earth System, which involves the individual and combined efforts of thousands of scientists from most of the world's nations, is an outgrowth of two long-term studies: the International Geosphere-Biosphere Programme (IGBP) and the National Aeronautics and Space Administration (NASA) focus on Earth System Science. In the United States, the $1.8 billion (annual) U.S. Global Change Research Program directs the scientific research of nine federal agencies, which, in turn, provide financial support to federal, academic, and commercial scientists in various research centers.

The question of the natural variability of the Earth System is so important because of the growing concern and mounting evidence that the exploding human population is having an increasingly negative impact on the Earth System. This negative impact especially affects the terrestrial and marine ecosystems that support an estimated 30 million to perhaps as many as 100 million species of plants, animals, fungi, and microorganisms.

Peter Raven notes:

> The atmosphere, soils, and waters are all conditioned by the existence of billions of years of life on Earth, and that life has in turn adapted to them. At present, though, human beings, having grown from 2.5 billion to 6 billion people in the short space of 50 years, are consuming or wasting about half of total net biological productivity on land, and about half of the total freshwater supply...leav[ing] little room for other organisms. Consequently, about [one-] quarter of the total [number of animal and plant species] is threatened by extinction within the next quarter century, and perhaps three quarters by the end of the century.[1]

There is no question that the twenty-first century will be a pivotal one in the future of Earth's biosphere, especially the future of humans. What will ultimately be the magnitude of the sixth extinction? Will the sixth extinction rival the past five mass extinctions of species that are documented in the geologic record? Will it perhaps even exceed the great mass-dying at the end of the Permian period 245 million years ago, when 50 percent of marine families became extinct?

The sixth extinction is fundamentally different from the previous five mass extinctions. This is because one species (*Homo sapiens*) is the causal agent, a species that has the scientific knowledge of its activities and a moral and ethical understanding of the consequences of these activities to the Earth System. Humans also have the decision making and the technological capability to stop and reverse the individual and collective impact of humans on the Earth, a process Garrett Hardin referred to as *The Tragedy of the Commons* (Hardin, 1968). In a later book, Hardin expanded upon this recurring theme and upon the importance of living within the physical limitations of Earth's finite environment. Armed with knowledge of the Earth Narrative, will humans, individually and collectively, act in the best interests of the Earth System? Or will they continue to ignore the compelling scientific story of the natural and human history of the Earth System and continue to degrade Earth's environment? The fate of the biosphere rests upon humans overcoming self-interest and upon embracing a planetary-wide conservation ethic.

4.6 THE IMPORTANCE OF THE EARTH NARRATIVE

Widespread acceptance of the Earth Narrative is critically important, because it provides *Homo sapiens*, the most intelligent species yet to evolve on Earth, with an objective understanding of their place on the planet and in the cosmos. The Earth Narrative also provides a clear understanding of the increasingly negative impact of human population growth and its associated activities on the Earth System, especially on Earth's biosphere. Earth's story reveals the scientific knowledge that can be used to persuade individuals, corporations, and nations to take the needed steps to halt and reverse the continuing degradation of Earth's numerous ecosystems and the accompanying and accelerating extinction of species. Many of the world's bioscientists and geoscientists are endeavoring to sound the alarm and establish a new conservation ethic; not the traditional approach to conservation (for example, protection of individual species or specific geographic areas or regions), but conservation from a planet-wide or Earth-System perspective. The title of a recent paper by Norman Myers, *What We Must Do to Counter the Biotic Holocaust*, is an example of one scientist's attempt to raise collective human awareness of our impact on the biosphere (Myers, 1999). The need for acceptance of the Earth Narrative by *all* people as a necessary prerequisite for creating a new global conservation ethic, based on *Do No Harm*. It is in the self-interest and collective interest of all individuals to accept this Earth Narrative, but it will take an enormous educational effort to accomplish the goal. The fate of the biosphere rests on the outcome of such an effort. Ideas about oneself, Earth, and the universe may have a rational (scientific) or a nonrational basis. The scientific Earth Narrative is based on reason and logic, not faith or myth.

4.7 IMPEDIMENTS TO THE ACCEPTANCE OF THE EARTH NARRATIVE

The Earth Narrative attempts to describe the place of *Homo sapiens* in the Earth System and in the cosmos, of which Planet Earth is but one tiny part. Contrary to religious dogma, nothing is special about humans other than the uniqueness of their complex brains and high intelligence with respect to all other life-forms on the planet. Because of their complex brains, humans differ greatly from other animals in many other traits, including anatomy, relative brain size, cognition, language and speech, culture, inventiveness, creativity, writing, artistic qualities, self-awareness, and sense of time.

Now, almost 5 billion years after Earth was formed from an interstellar gaseous nebula that give rise to the Solar System, and about 3.8 billion years since single-cell prokaryotic life first arose on the planet, humans are but one of 1.7 million species formally classified by biologists—out of an estimated total of 30 million to 100 million. Humans are the most recent hominid of at least 11 species of biped primate mammals to have evolved during the past 5 million years, ever since the hominid evolutionary line parted from the other African great apes. The DNA of *Homo sapiens* differs genetically from its closest living relative, the chimpanzee, by less than 2 percent. But this 2 percent makes human beings unique among all primates and among Earth's other 4,500 species of mammals.

It is important for humans to realize that we are but one of the 4,500 mammal species that exist on Earth and that we are but one species of tens of millions that form the biosphere. We also need to remember that all species exist within the physical limits imposed by the geosphere and that all species live on energy from the sun. We are totally dependent on the biosphere for food, oxygen, and other necessities of life. Humans are one of the evolutionary results of biological evolution operating over the lengths of Geologic Time. We find ourselves on Earth, after almost 5 billion years of Earth history, by chance, not by design.

Widespread acceptance of the new Earth Narrative will be difficult to accomplish because myths, especially those that are part of a religion, can hold enormous power over how human beings view themselves, Earth, and the cosmos. Myths have an extraordinary influence on the beliefs and ideas of most people in most cultures and religions. Religion is defined in Webster's dictionary as "the service and worship of God or the supernatural...; a personal set or institutionalized system of religious attitudes, beliefs, and practices...; a cause, principle, or system of beliefs held to with ardor and faith," past and present. The late Carl Sagan, one of the prominent scientific minds of the twentieth century and a strong advocate for greater public understanding of science and of the place of human beings in the cosmos, worked much of his life to help dispel myths that block such understanding. He looked on "science as a candle in the dark" in a "demon-haunted world" (Sagan, 1997b).

After Charles Darwin published *The Origin of Species* in 1859, he was attacked by religious leaders and others of his time. Christian fundamentalists continue to hinder the teaching of biological evolution and the history of Earth in some school systems, particularly in the United States, even today. Darwin's staunchest defender was Thomas H. Huxley. Today that role has fallen on the shoulders of writers and scholars such as the Harvard University geologist and historian of science Stephen Jay Gould (1999) and the paleontologist Niles Eldredge (1995). The National Academy of Sciences (1998, 1999) and many national scientific societies such as the American Association for the Advancement of Science and the National Association of Biology Teachers, as well as natural history museums and other national institutions, are also making great contributions. That the United States, with a seemingly excellent system of education, has been unable to separate religious mythology from scientific facts, both in the science curricula and in the biological and Earth sciences texts used by numerous schools, is a testament to the enduring power of myths and the zealotry of groups that embrace such beliefs. The influence of these groups limits the quality of science education in many school districts in the United States, either directly through local school boards, or indirectly by the limitation in biology and Earth-science textbooks.

These anti-science efforts are especially destructive to the development of independent thinkers among our young people. For science to move forward in a science-based technological society such as the United States, the nation must be able to use all of its available brain power. Creativity and innovation spring from individual human brains; no society can afford to lose any of its intellectual potential.

Failing to provide an adequate scientific education, especially in the Earth and biological sciences, to large numbers of students in the United States also causes a significant negative impact on national and international conservation policy. This is a tragedy whether the result is a negative impact on funding for global environmental change research, on support for preservation of biodiversity, or on funding for ways to limit human population growth. For example, Manuel Lujan, Secretary of the U.S. Department of the Interior between 1989 and 1993, believed that Earth was only 6,000 years old. The Secretary of the Interior's acceptance of a religion-based myth, a seemingly harmless personal religious belief, resulted in the Department of the Interior's *not* being a significant contributor to the U.S. Global Change Research Program. This is an example where ignorance of the natural and human history of Earth on the part of individuals can have detrimental and far-reaching public-policy ramifications.

As Niles Eldredge noted in his aptly titled book, *Dominion*, humans "have become the first species on Earth to interact as a whole with the global system" (Eldredge, 1995, p. xv). The title of Eldredge's book is taken from Genesis: "And God blessed them, and God said unto them: Be fruitful, and multiply, and replenish the earth, and subdue it: and have dominion over the fish of the sea, and over the fowl of the air, and over every living thing that moveth upon the earth." To fundamentalist Christians who believe in a literal interpretation of the Bible, this is a powerful command. These words, sacred to many believers, Christians and others, were written down when the global population of Earth was about 300 million, and the true nature of the biosphere, with its 30 million or more species, of which humans were but one, was unknown. Not only does this command order humans to reproduce in numbers sufficient to subdue the geosphere and biosphere of the Earth System, but it orders humans to have dominion over all life-forms on Earth—the biosphere. It is estimated that 76 billion human beings have already lived on Earth during human history, and the expansion of the human population to 6 billion at present and to 12 billion by 2050 gives every indication that humans will satisfy or achieve this environmentally insensitive biblical command. The impact on Earth's environment of uncontrolled population growth and the desire of people to have supreme authority and absolute ownership of all living things show a lack of understanding as to the ramifications of such a population to the Earth System. In addition, it says that humans are set apart from all other life-forms in the biosphere, another myth that is unsupported by the factual evidence of biological evolution. As Lynn White stated in his essay "The Historical Roots of Our Ecological Crisis," "Especially in its western form, Christianity is the most anthropocentric religion the world has seen" (White, 1967).

There are, of course, scientists and religious leaders who understand both the power that religion has to guide humans in their individual behavior and collective activities and the need to join together and take action to preserve Earth's environment. In January 1990, Carl Sagan, Stephen Jay Gould, Lynn Margulis, and Peter H. Raven were among 32 leading scientists from all over the world who wrote an appeal to the world's religious leaders for a joint commitment on the environment. Within six months, 270 religious leaders supported the appeal, which gives some hope that the scientific Earth Narrative can be accepted and that actions can be taken to preserve Earth's biodiversity. Three excerpts from the joint commitment, "Preserving and Cherishing the Earth," are instructive:

> ...a voluntary halt to world population growth—without which many of the other approaches to preserve the environment will be nullified..., the environment[al] crisis requires radical changes not only in public policy, but also in individual behavior.... At the same time, a much wider and deeper understanding of science and technology is needed.

We have a great need to establish a new conservation ethic that includes the *entire* Earth System. The inclusion of scientists, religious and political leaders, policymakers, and others in this goal is crucial for maintaining natural processes and rates of change that preserve the natural biodiversity of the planet.

This monumental effort will be primarily educational so as to increase public understanding of science and to stress the importance of both the scientific method and continued natural-science studies in order to add more details to the Earth Narrative. We must accomplish this goal in the face of determined opposition from religious fundamentalists and anti-modernists. Through education, understanding, and cooperation a consensus must be reached to accept and promote this vital perspective of a scientific Earth Narrative, and to establish a global conservation ethic of *Do No Harm*.

It is no accident that field biologists, geologists, astronomers, and anthropologists, those most intimately involved in studies of the natural and human history of Earth and in investigations of the human impact on the Earth System, have been most passionate in writing books and articles directed to nonscientists. Rachel Carson, Carl Sagan, Stephen Jay Gould, Niles Eldredge, James Lovelock, Ernst Mayr, and Edward O. Wilson have been at the forefront of these efforts to educate nonscientists on the concept of biological evolution and on the magnitude of the impact of humans on the planet. Other scientists, such as Loren Eiseley, and science writers such as John McPhee have published important series of books on the natural history of Earth.

One example of the effect scientists can have is the work of Rachel Carson. Carson, a biologist and prize-winning author, wrote several books on natural history, including *The Sea Around Us*. In 1962 she published a book that would change forever our perspective on the human impact on Earth's environment by alerting us to the danger of chemicals, especially the widely used pesticides, to the fabric of life on the planet. *Silent Spring* helped launch the environmental movement and forced the U.S. government to address seriously the impact of artificial chemicals on all the biota in the biosphere, including humans.

4.8 CONCLUSIONS

It is without doubt that the actions that human beings take, individually and collectively, during the twenty-first century will determine the fate of many millions of species, including our own. Aside from ethical implications of causing the extinction of a large percentage of Earth's biota, it is imperative that we address the human-population growth problem quickly—to do so is in our own self-interest.

According to David Pimentel, more than 3 billion humans are now malnourished, one-half of the global population in 1999. By comparison, 2 billion was the total world population in the 1930s, only 70 years ago (Pimentel et al., 1999). The same authors conclude that additional growth of the human population will eventually confront the limitations of food production on degraded land, the degraded quality and finite quantity of available water resources, and the increased incidence of disease and contaminated environments, both in terrestrial and marine ecosystems. Humans must live within limits that are controlled ultimately by the finite natural resources of the planet.

The natural and human history of Easter Island is, in many ways, an Earth Narrative in a microcosm. Once a forested island, 166 km^2 in area, 3,220 km west of South America, it was first settled by Polynesians in about A.D. 400. The population grew rapidly from a few to 7,000, or perhaps as high as 20,000. Within 1,000 years of settlement the human population consumed all its terrestrial natural resources, which wiped out the forest and caused most of its plants and animals to become extinct. The society then disintegrated, so that by 1700 the human population declined to 10 to 25 percent of its former number. Will the fate of the Easter Island ecosystem eventually become the fate of ecosystems on Earth's largest "islands," its continents?

Humans, as a species, have become so successful at modifying their environment that many of the natural limitations on the expansion of populations of our fellow animals have been overcome by technological innovations. As Peter Raven noted, humans, at a current population of 6 billion, are consuming or wasting about 50 percent of the total net biological productivity on land and 50 percent of the available supply of freshwater. The overwhelming and expanding human presence leaves less and less room in the environment for other biota. Either the physical and natural resource limitations of our planet or emergent diseases could force an upper limit to human population, but technological advances may continue to raise that projected limit. The ultimate loser will be Earth's biosphere, as well as the quality of human life.

The failure of humans to comprehend and understand the Earth Narrative, the scientific story of the natural and human history of Earth, especially the place of humans in it and their accelerating negative impact on the Earth System, will lead to dire consequences for Earth's biosphere. Logic and reason must prevail, if we are to survive as a species. We must establish an Earth-System-based conservation perspective that is focused on the complete preservation of Earth's biotic inheritance and that is based on a *Do No Harm* ethic.

We see glimmers of hope, even during a global population increase of 90 million annually (a net of three more people are added each second: nine humans are born and three die every two seconds). As Norman Myers noted, a marked increase in environmental awareness is evident on a global basis. All field geoscientists and bioscientists and others who study some aspect of the Earth System must become teachers and devote some of their time both to increasing public understanding of the Earth Narrative and to counter what Myers terms "the biotic holocaust."

I would like to close this essay on a positive note by citing two success stories in North America in the conservation of Earth's biota. The American bison once ranged from Oregon to New York and from the Northwest Territories of Canada to México. Millions of bison grazed the central plains of North America. By the late 1800s, their numbers were reduced by hunters to only a few hundred. Protected herds in limited areas in the United States and Canada now number about 200,000.

By 1912, trumpeter swans were extirpated from their natural areas by hunters and egg gatherers. With fewer than 100 swans known to exist (not including unknown separate populations in Alaska) conservationists worked for their protection through national protection laws and for the designation, in 1935, by the federal government of 22,000 acres of nesting habitat in the American West. Several thousand trumpeter swans are now distributed seasonally in the northwestern United States, western Canada, and Alaska, and efforts are now in progress to restore them to their former East Coast ranges.

In both of the examples given, education followed by specific actions, such as legislation, strict protection, and dedicated activism, saved two species from extinction. We no longer have time, however, to rescue species one by one. A conservation ethic that is planet-wide in scope must quickly emerge.

ACKNOWLEDGMENTS

The breadth of the subject matter covered in this essay required reviews of the text by many other scientists. I want particularly to acknowledge the thoughtful and comprehensive reviews by Charles H. Southwick, Department of EPO Biology, University of Colorado at Boulder; C. Wylie Poag, Woods Hole Field Center, U.S. Geological Survey; David Pimentel, Department of Entomology, Cornell University; Hans-Dieter Sues, Department of Palaeobiology, Royal Ontario Museum; John C. Steele, Marine Policy Center, Woods Hole Oceanographic Institution; Christopher D. Williams, Division of Reproductive Endocrinology and Infertility, the University of North Carolina at Chapel Hill; George E. Watson, retired Curator of Birds, Smithsonian Institution; and Thomas C. Aldrich, Woods Hole Field Center, U.S. Geological Survey.

I am also indebted to the superb copyediting of the manuscript by Susan Tufts-Moore. I want to acknowledge the fine job done in typing and page layout by my most capable senior administrative assistant, Janice G. Goodell. Preparation of the two figures was done by Kirsten E. Cooke, computer graphics specialist par excellence.

NOTES

1. Peter H. Raven, written communication, August 7, 1998; from his lecture on "Biodiversity, the Global Environment, and the New Millennium," Marine Biological Laboratory, Woods Hole, Massachusetts.

REFERENCES

Diamond, J. 1997. *Guns, Germs, and Steel.* New York: W. W. Norton & Co.

Eldredge, N. 1995. *Dominion.* New York: Henry Holt.

Gould, S. J. 1999. *Rocks of Ages—Science and Religion in the Fullness of Life.* New York: Ballantine.

Hardin, G. 1968. "The tragedy of the commons." *Science, 162* (3859), pp. 1243–1248.

Leakey, R., and Lewin, R. 1995. *The Sixth Extinction—Patterns of Life and the Future of Humankind.* New York: Doubleday.

Myers, N. 1999. "What We Must Do to Counter the Biotic Holocaust." *International Wildlife, 29* (2), pp. 30–39.

National Academy of Sciences. 1998. *Teaching About Evolution and the Nature of Science.* Washington, DC: National Academy Press.

National Academy of Sciences. 1999. *Science and Creationism—A View From the National Academy of Sciences* (2d ed.). Washington, DC: National Academy Press.

Pimentel, D., et al. 1999. "Will humans force 'nature' to control their numbers within the limits of the Earth's resources?" *Environment, Development, and Sustainability* (in press).

Sagan, C. 1997a. *Billions and Billions: Thoughts on Life and Death at the Brink of the Millennium.* New York: Random House.

Sagan, C. 1997b. *The Demon-Haunted World: Science as a Candle in the Dark.* New York: Random House, 457 pp.

Shermer, M. 1997. *Why People Believe Weird Things: Pseudoscience, Superstition, and Other Confusions of Our Time.* New York: W.H. Freeman.

Swerdlow, J. L. 1999. "Biodiversity: Taking Stock of Life." *National Geographic, 195* (2), pp. 2–5.

White, L. 1967. "The historical roots of our ecological crisis." *Science, 155* (3767), pp. 1203–1207. [Also *in* Detwyler, T. R. 1971. *Man's Impact on the Environment.* New York: McGraw-Hill, pp. 27–35.]

5

Messages in Stone: Field Geology in the American West

Christine Turner

Christine Turner is a senior research scientist at the U.S. Geological Survey (USGS) in Denver, Colorado. During her two decades of service for the USGS Turner has held several management positions, and has also chaired a national committee on the future of the USGS. She is best known for her reinterpretation of the formation of a class of uranium deposits, in which she overturned previous models. Her current research focuses on two topics: the reconstruction of Mesozoic era landscapes in the American West in order to describe the ecosystem inhabited by Jurassic dinosaurs, and the development of ways to better integrate the sciences and the humanities for the resolution of community land use and resource controversies.

In this essay, Turner offers a personal account of the nature of geologic fieldwork. In this account, field studies provides much more than sets of data. Fieldwork also exemplifies the creative and imaginative nature of human rationality, offering a kind of "rock logic" distinct from the rationality characteristic of the lab sciences. Field reasoning offers a more realistic account of the nature and limitations of scientific reasoning, and indeed of human reasoning in general. This last point is particularly important in a time when the public looks to science to provide answers to questions of common interest.

Turner is concerned with the decline of field-based scientific research within the USGS and elsewhere, and the concomitant loss of a specific style of thinking within our culture. In the rush to embrace the computer revolution we have lost sight of the need for slow, deliberative thinking, and for contact with nature and the real things in the world.

At El Morro National Monument in New Mexico, a wall of rock surrounds a pool of water that has served as a watering place for travelers and explorers who have passed this way over the centuries. Peering into the quiet depths of the pool, one can imagine weary wayfarers kneeling to slake their thirst after crossing the desert that surrounds this oasis. Some recorded their passage, etching their messages in the sandstone cliff beside the water hole. The Ancient Ones carved their mysterious petroglyphs on the rock walls. Centuries later, Conquistadores passed by, searching for new souls to convert to their God and to bring under the dominion of Spain. Other passersby were part of the camel train led by Colonel Beale, who thought these animals better suited to the arid West than horses.

The Anasazi petroglyphs date back to the thirteenth century, and the Spanish inscriptions carry dates as early as 1604. These inscriptions have faded, abraded by wind and water and sand swirling around the desert landscape; yet their messages are still decipherable. One can only wonder why their authors took such care to tell us of their passage, gracefully inscribing names and symbols into the rock. They would have many stories to tell.

Gazing at the cliff face at El Morro, trying to envision the past world that these early explorers inhabited, I realize that interpreting their engraved messages is similar to the work I do as a field geologist. The geologist deciphers the stories etched in ancient rocks from just such faded but tantalizing clues. The secrets of the geologic past lie quietly entombed in cliffs of rock: The geologist spends days passing from outcrop to outcrop, cliff to cliff, canyon to canyon, hoping that the messages will become clearer as more pieces of the story are gathered. The geologist's search for understanding even carries into evening reveries in camp, especially when the same cliffs that were climbed during the day are now faintly visible by the glow of the moon.

As a field geologist with the U.S. Geological Survey in Denver, Colorado, I have studied sedimentary rocks on the Colorado Plateau for 24 years. I have worked down side canyons that lead into the Grand Canyon of Arizona, pulled myself up steep cliffs in northern New Mexico, and walked along mesas in western Colorado and down gorges in eastern Utah. Each time I went in search of a story: a tale of ancient seas and sand dunes in the Grand Canyon, of uranium deposits in former stream beds that now form red rock cliffs in New Mexico, of former alkaline lakes that stretched across much of western Colorado, and of dinosaurs who lumbered across the landscape and died on the ancient plains of Utah. Each time I had to find the right questions to ask, and search patiently for the answers. The rocks have revealed their stories much as a master storyteller around the campfire teasingly builds suspense before parting the curtains of mystery.

A field geologist moves with the rhythms and vicissitudes of nature and learned the art of patience. During the course of a summer spent searching for clues to the story that we know awaits us somewhere in those cliffs, we must deal with life as it comes, not as we wish it would be. A field geologist cannot force the answer from the rocks any more than he or she can command suitable weather for the search. In a life spent outdoors, setbacks that would never be tolerated in the comfortable ambiance of home are taken in stride. In the field, the world of the real starts to transform what we might otherwise consider the "real world," and the realities of nature irrepressibly assert themselves over the perceptions of reality born of too many months in the office. One learns the lessons of nature slowly, through the daily labor of fieldwork among the majestic mesas and colorful canyons. These experiences inform and shape a person, making one more attuned to nature's whims, beauty, and wisdom.

Invariably, I feel a letdown every autumn when I return to the office and exchange the real world of nature for the confines of a concrete building. The broad perspectives so slowly won beneath a hot sun are often shattered under the glare of fluorescent lights. When I first traveled into the open lands as a field geologist, I held the lessons of stone and wind and desert skies quietly inside, enjoying the privacy of thought and movement and understanding that draws so many of us to the profession. But with the coming of the new century it has become apparent that geologic facts and perspectives will be crucial for confronting the challenges we will face. It is time for field geologists to share their stories with those who might be invigorated by the lessons of life lived through and with the earth.

5.1 DOING FIELD GEOLOGY

Like most field geologists, I often take a student with me as a summer field assistant, usually a student who has completed his undergraduate training and has taken field camp, a sort of "boot camp"-type course that exposes students to the rigors of field geology. I do this because the work requires physical labor and because there is safety afforded by a companion in remote places. The student packs his duffle bags, sleeping bag, and enough clothes and sunscreen to survive the prickly vegetation and the fierce sun that awaits him on the Colorado Plateau. On the drive from Denver west toward the Colorado Plateau, the hours on the road are often spent answering questions that the eager neophyte has about the adventure awaiting him. One particular question, asked by Doug Boyer, a geology major from Penn State University, fascinated me: "How do you 'do' geology?" He reached back and pulled a dog-eared article from his duffle bag—it was T.C. Chamberlain's 1897 article on the scientific method. Doug was fascinated not just by the facts that he had learned from his undergraduate training, but by the methodology, the ways that geologists actually do their work. He recognized that something more was required than the techniques he had learned in his field-methods class. This question sparked a summer-long conversation about how to "do" geology.

As in many professions, it is more often in the "doing" that true learning takes place. As with all sciences, geology has its textbooks and procedures for how to interpret the rocks, and field equipment for taking certain types of measurements and for gathering data. A key characteristic of geology, however, is that we must interpret environments that have long since vanished. We must reconstruct the past from fragmentary evidence found within those strata. As an historical science, geology requires that we learn the skills of Sherlock Holmes, working with highly circumstantial evidence and hoping to convince the jury of our scientific peers that we have enough data to seek a verdict.

There is no way to return to the scene of the crime, however, because we are unraveling mysteries that occurred many millions of years ago. Nature has had ample time to perform a massive cover-up operation. Evidence has been modified by subsequent events or has been removed altogether. But as geologists we learn quickly to deal with incomplete data and with enormous uncertainty, and become so accustomed to extrapolating from fragmentary data that we forget how different our science is from chemistry and physics, where carefully designed and reproducible experiments are the final arbiter of theories.

Working intimately with "deep time," we have developed a repertoire based upon comparative approaches. We infer that "the present is the key to the past," and therefore assume that streams that deposited sand and mud a million years ago probably had the same hydraulic properties that characterize modern streams. But what we do not know for any particular ancient stream are all the particular variables that affect stream behavior, such as gradient, amount of vegetation, climate changes, and seasonality of precipitation, just to name a few. Deciphering rocks is an uncontrolled experiment: We cannot hold all of the variables constant and vary just one at a time. We can never know all of the factors that combined to produce the resulting strata—we only have the result and reasonable inferences gained from analogy with modern processes.

Perhaps the most important ability of the field geologist is the capacity to envision and even to "inhabit" ancient landscapes by way of deliberate day dreaming. A geologist climbs into an imaginary time machine, where every observation, every cryptic piece of evidence in the rocks is viewed not as part of a stonily silent cliff but as a landscape as vibrant as it was at the time the sediments were deposited. When a geologist finds a trace fossil, formed by a disruption of sediments when an organism burrowed into a river sand a hundred million years ago, he can imagine being that organism. He hears the waves crashing overhead, and imagines burrowing deeper. Strata left by an ancient stream migrating across a sandbar pulls the geologist back to a river 150 million years ago. He can envision gravel tumbling along the bar during flood stage, stinging his feet, his boots providing no protection against the deluge. Perhaps Sarah Andrews (1998), a petroleum geologist and writer of mysteries with geologic themes, says it best: "Geologists have the ability to see into solid rock and imagine entire worlds."

Because geological evidence is so fragmentary, what counts as "knowing" for geologists is very different from what might be considered statistically reliable in other sciences, such as biology. Thure Cerling, a paleoclimatologist at the University of Utah, has worked with biologists who, in an effort to understand evapotranspiration rates in a modern ecosystem, will sample a thousand leaves from a single tree and determine the isotopic composition for each leaf—an abundance of data undreamed of in field geology. The biologist can note the amount of sun exposure for each leaf during the day, and thus the evapotranspiration rate for each leaf, which is factored into isotopic interpretations. Cerling noted that we geologists consider ourselves lucky if we find one carbonized impression of a leaf in an interval of rock that may have been deposited over a period of 10 million years, and then we extrapolate the results to the entire 10 million-year time period!

Geology often is not as precise as the more analytical side of our natures might wish it could be. In fact, geology is inherently imprecise. This is not a discipline for those who like their numbers with decimal points. A case in point is the measurement of the orientation of a fault plane. Faults by nature define curving, undulating surfaces as they cut through the rocks, and therefore precise determination of the orientation and angle of dip or "attitude" of the fault plane is not possible. A field geologist may be able to measure a fault plane where it intersects Earth's surface (if vegetation does not get in the way!), and can obtain indirect measurements of the same fault as it projects into the subsurface by looking at seismic lines that model the layers of rock in the subsurface. Nevertheless, chances are that no two measurements will be identical. Deciding which is the "right" one is a judgment call. Geology's reliance on judgment calls would drive some in other disciplines crazy, but for one blessed (or cursed!) with a geologic mentality honing one's judgment to a fine edge becomes a source of great pleasure. In any case, the general trend of a fault is sufficient for most geologic studies, and measurements are routinely taken with the understanding that the measurements are accurate for where they are taken along the fault trace.

At Yucca Mountain, Nevada, which scientists are currently evaluating as the potential repository for the nation's nuclear wastes, concern about the trend of faults is a factor in the site evaluation. Engineers concerned about predictability sometimes request more precise measurements and predictions than the geologic data permit. Whereas the inherent imprecision in many geological measurements is viewed by some as a deficiency, geologists merely see the imprecision as a reflection of the inherent vagaries of real-world phenomena. Geologists like to be accurate in their measurements, but they recognize that mathematical precision or certainty is often an unrealistic expectation.

Perhaps the most important aspect of geology, and one that significantly constrains our interpretations, is its integrative nature. Our goal is to develop an interpretation or synthesis of the data that best accommodates the disparate kinds of data that we obtain. We complement our field observations, made with the "scanning geological eyeball"—our most valuable piece of field equipment—with laboratory analyses, and then work to accommodate and combine the implications of each data set. This holistic sort of reasoning helps to overcome the inadequacies of any particular set of data. With each observation or analysis we reevaluate the entire story in light of the new evidence. We do not rest until we believe we have come up with the interpretation that best fits the data and honors the observations. Interpretations in geology are made at different scales, and from different types of data. Observations and interpretations are constantly reevaluated with respect to the developing picture until an integrated story emerges, rich in detail but richer still in its wholeness.

The entire brain comes into play in the "doing" of geology, for we are constantly searching for order in the world (and in ancient worlds!). The search for order is largely a left-brain exercise, whereas making observations is the domain of the "visual" right brain. A small war is constantly waged between the two hemispheres of the brain as we do field geology. Our left brain brings to the outcrop all of the latest models and interpretations from the scientific literature, ready to overlay the "template of truth" on every observation we make. The left brain is happiest when it has an answer, and is quite satisfied to abide by conventional wisdom. The right brain is the visual hemisphere and sees the world in its variability. If the right brain sees something anomalous that does not fit the preconceived notions of the left brain, it demands a new explanation. The desire for order in the human psyche sometimes overrides new contradictory observations.

Creativity in geology is the ability to let the "signs" in the rocks tell their story, and to interpret them with the aid of current hypotheses. The understandings in the left brain are gained by one's own experience as well as the experience and beliefs of other geologists, learned from group field excursions and the geologic literature. The new observations are constantly evaluated by what one already "knows." It is sometimes difficult to let the right brain work unfettered by the knowledgeable but sometimes overbearing left brain, which seeks to categorize everything by what it already knows. Models and hypotheses are essential to this process, because they ensure that our observations do not remain random. They also provide a framework within which to systematically order geologic observations.

Breakthroughs and new ideas in geology come about when geologists are receptive to the possibility that the rocks might be trying to tell them something that cannot be reconciled with preconceived notions. Most new ideas, once adopted, seem apparent to the next generation of geologists, because the new understandings become part of the list of possible answers brought to each new situation. It is easier to follow in the wagon ruts of conventional wisdom than it is to strike out on new paths. Creativity in geology is possible when we "see" something as if for the first time, and allow ourselves to be gently guided but not coerced by our knowledge and experience. In essence, the rocks know what happened to them. It is our responsibility to see them as they are and listen to what they are trying to tell us, since they will eventually reveal their stories to us.

5.2 GEOLOGY AT THE CROSSROADS

Although a fundamental goal of geologists is to decipher sequences of geologic events or processes, we are now being asked by society to interpret Earth in a manner that focuses more on the modern landscape and in ways that directly affect humankind and our fellow inhabitants of the planet. Although geologic stories of worlds past stir the imagination and provide valuable insights, an urgency exists to confront major societal issues as we enter the next century. Such issues as global warming, preservation of wetlands, concentration of population in areas prone to natural hazards, and concerns over the quality of the air we breathe and the water we drink will require the special perspectives and understandings of geology. The interconnectedness of all natural systems is now widely realized and requires the integrative approach that characterizes geology. Geologists understand that the world is complex, with problems that do not lend themselves to precise answers. Geologists are therefore well suited to a world that is characterized by imprecision and uncertainty.

But just as geology is poised to take its place at center stage to help society decide how to respond to major societal problems, we find geologists at a technological and psychological crossroads. Increasingly, field geologists are putting away their rock hammers and relinquishing their field vehicles, believing that field studies will no longer be part of their research. The same geologists who once scrambled over outcrops in search of subtle clues to a geologic mystery are joining the ranks of those who spend their days in front of a computer screen and engage in hallway conversations about the latest software or upgrade.

One could argue that this is the age of the computer and that geologists are merely embracing the new technology. Indeed, computers are opening the way to new understandings of Earth. The computer enables the handling and manipulation of enormous data sets. Forward modeling of complex systems and the ability to monitor Earth processes in "real time" are revolutionizing our science. The move to computers can be viewed as a gain in sophistication for our science.

A mindset accompanies this trend, however, and it is the mindset, rather than the use of computers, that is troubling. Computers have raised our collective expectations in ways that have unexpectedly negative consequences. Now there is little excuse not to generate vast data sets, create elegant computer models, and respond instantaneously to an avalanche of E-mail. Geologic research is being overtaken by the same kind of frenetic activity, substitution of technology for reality, joyless self-flogging, and disconnectedness from nature that characterizes most modern work-places. We are joining the trend to abandon the natural for the seemingly greater seductions of technology.

Thoughtful reflection and deliberation are being replaced by hyperactivity in every aspect of our lives. We find ourselves feeling guilty for somehow not "keeping up," even when to do so would be bad for our health and sanity, as well as for truly accomplishing the work we originally set out to do. Most discouraging, however, is that the sense of the "real," has been replaced in our modern culture by the "hyperreal" (Borgmann 1993). The hyperreal is exemplified by an IMAX theater at the entrance to Zion National Park. A sign outside the theater urges tourists to skip the park entirely and save time by seeing the movie instead. This way, a tourist on a hectic "vacation" schedule that allows little time for meaningful experiences can easily make Las Vegas by nightfall, perfecting the art of remaining totally detached from natural experiences even when on a trip whose original goal was to experience nature! The IMAX movie and Las Vegas both exemplify the hyperreal—a reality both more stimulating and shallow than the world of nature.

A sculpture commissioned for the U.S. Geological Survey (USGS) headquarters in Reston, Virginia, symbolizes the invasion of the hyperreal into an agency that stands as one of the last bastions of the real. The sculpture consists of a tree stump and adjacent boulders, all covered entirely in steely metallic paint. The sculpture stands in stark contrast to the naturally sculptured columnar basalt that graces the entrance to the USGS grounds. The addition of the metallic sculpture is a dissonant symbol of the way the Earth sciences might go—the betrayal of the natural for the glitter of the hyperreal. Geologists who work in the building could not resist placing a plastic flamingo near the metallic sculpture in mockery and protest, adding aluminum foil eggs to the scene each spring.

Disconnection from the Earth we inhabit, the increasing substitution of the hyperreal for the real, and the replacing of reflective thought with frenetic hyperactivity, are insidious and major threats to our well-being and to our society. The field geologist who remains grounded in the real has a connection to Earth that is becoming increasingly tenuous, just as society's connection to the real is becoming increasingly tenuous.

Many field geologists still have a remembrance of the vagaries of nature and the elusiveness of control. We have still fresh memories of the joy of the natural, and the calmness that leads to creativity. Most importantly, field geologists know about being connected to the very Earth they are trying to understand. They are grounded in ways that are increasingly unfamiliar to modern society. These are valuable perspectives that could be shared with society but, ironically, field geologists are losing those very perspectives just when they are most sorely needed.

5.3 BEING A FIELD GEOLOGIST

Field geologists' perspectives about life and their intimate connections with Earth are more valuable to contemporary culture than any information provided from specific geologic studies. Beyond the importance of the unique approaches of geologic inquiry are the unique perspectives that a life in the field engenders: "Being" a field geologist informs as much as "doing" field geology. In the modern workplace most of us have little time to "be" because we are so busy "doing."

Organizational consultants tell us that all of our daily work activities can be divided into four categories or quadrants (Covey, 1989). The first quadrant includes all of the important and urgent activities that come across our desks. Budget deadlines would fall under this urgent quadrant, for without the budget the work cannot get done, so it qualifies as both important and urgent. The second quadrant involves all the important but nonurgent activities, such as long-range planning, achieving the goals of the organization, and other activities that may have no deadline but are the most important things we do. The third quadrant is the category saved for all of those urgent but unimportant things that pile up on our desks and require attention, usually because someone else in the organization has imposed a deadline for their completion even though they are not important in the grander scheme of things. The fourth quadrant is reserved for activities that are not important and not urgent.

In the various management training classes I have had, it was noted that we spend most of our time between activities that fall in the truly urgent first quadrant, and the third quadrant, where we take care of the paperwork that piles up on our desks or the little chores that arise. That is to say, we respond to the important crises, and then return to our desks and try to clear it of all of the trivial but seemingly necessary details, just in time to be pulled off again for another crisis. Getting to the second quadrant, where the important work of the organization takes place, is the goal, but is often difficult. Fourth-quadrant items should be crossed off our lists, say the organizational consultants, because if the activities are not important and not urgent, why would we do them at all?

But there is another way to conceive the four quadrants, in which the fourth quadrant is seen as the crucial one. You get to the second quadrant—the important things, which take reflective thought—by way of the fourth quadrant. The fourth quadrant is where you take a walk, play with the children, do "nothing much" by contemporary standards; and in that relaxation, you find the reflectivity and the calmness that you need to enter second-quadrant activities. Fieldwork exemplifies the combination of second-quadrant and fourth-quadrant activities. First- and third-quadrant activities are left at the office, and the work of doing geology begins for field geologists when they head out into the open air. The reflective moments at sunset, lunchtime, or sitting around the campfire are as essential to the work as the actual labor performed at the outcrop. The mind has incubation time for thoughts and ideas. It is through the inhabitation of such reflective spaces that creative work is done.

It has been said that "creativity requires the encounter of the intensely conscious person with his or her world" (Wantland, 1998). Buddhists speak of being present in the moment, and in the field this seems to happen effortlessly. Fieldwork engages and enlivens all of the senses. One feels the earth beneath the feet and hears the gentle sigh of the breeze. The rocks are at once tactile and visual, with textures as important as visible features.

The senses are rarely assaulted in the field the way they are in the city. Ears that were accustomed to tuning out traffic noise and the cacophony of modern life reopen in the field to hear the more gentle, calming sounds of nature with their subtle messages. On many a hot summer day, there is only the whisper of a gentle breeze and the swooshing sound of a hawk circling overhead. This is not noise. These are sounds—sounds that relax and inform rather than tense the body.

So many pleasures that we normally take for granted are cherished and savored in the field. Any field geologist who has walked beyond the distance a canteen of water permits on a summer day in the desert fully appreciates that first swallow of water after returning to the truck where more water is stored. Setting up a folding chair under the reviving shade of a lone cottonwood tree at lunchtime on a hot summer day has the same effect.

The simplicity of life in the field is integral to the geologic experience. A new block of ice in the cooler, fresh water in all the water containers, enough charcoal for a barbecue in the evening, and it is possible to "dry camp" near the next day's outcrop. Camping close to "work" is often possible on the Colorado Plateau because of the small amount of land that is privately owned. Pitching the tent at the base of the outcrop allows a field geologist to work on the outcrop by day and muse about the outcrop with the setting sun, as the evening shadows play on the multihued layers of rock. Soon, the coyotes begin their evening song. It is a matter of efficiency to camp so close to the next day's work, but it is also a way of living with the outcrop, to be receptive to stories that emerge from the seemingly impassive cliffs. Inhabiting the modern landscape of the Colorado Plateau by camping in the moonlight is conducive to reconstructing the ancient landscapes. The Earth is more tangible when you sleep on it.

Shade, water, a bath taken with a bucket, food grilled over the barbecue, a folding chair, and a level tent site become the finer things in life. The shade seems cooler, the water more thirst-quenching, the bath more cleansing, the food tastier, the chair more comfortable, and the sleep more restful in the field than in any other place. The body seems more suited to the rhythm of field life than life back in the office.

Even the setbacks experienced by the field geologist are instructive. As might be expected in a life spent outdoors, the setbacks are often related to the weather. As a summer day begins on the Colorado Plateau, blue sky often extends from horizon to horizon. By mid-morning a wisp of cloud may appear in the otherwise blue sky, and, ever so imperceptibly in the course of the day, it grows into a thunderhead. The thunderstorms rumble across the region, and the "baggy pants" clouds develop streaks of rain beneath them. Field geologists watch the clouds, ever mindful of the danger of lightning and the possibility of flash floods, both hazards of working during monsoon season on the Colorado Plateau.

The vicissitudes of nature are taken in stride, and patience in the face of the inevitable setbacks becomes the normal response. Life is transformed into a balance of hard labor and soulful rest. Despite longer working hours, the weariness at the end of the day is physical rather than spiritual. The labor is hard, but satisfying. Physical as well as mental powers are fully engaged in the daily tasks. As the spirit calms, a sense of reverence for Earth grows, and one feels grounded rather than disconnected. And in turn, connection to the Earth engenders a receptivity to its stories.

One can also learn from other dwellers of the landscape—for instance, the Navajo people—Native Americans who live in the Four Corners region of the Colorado Plateau. During one research project, the rocks that we wished to study lay in tilted strata along the west side of the San Juan Basin in northwestern New Mexico, within lands of the Navajo Nation. The small community of Toadlena lay between the highway and the outcrop. Being mindful of the sensitivity of the Navajo to Anglos traipsing across their lands, and also being a little unsure of the route, we stopped beside a hogan that lay athwart our route. A Navajo came out of the hogan and stood silently while I explained where we wanted to go. There was no reply. Even a letter from the Tribal Council that granted permission to enter Navajo land elicited no response. The Navajo looked at the letter, but said nothing. It seemed appropriate to try explaining again what I wanted. Again, no answer, but polite attentiveness. It occurred to me that in some parts of the reservation the Navajo speak no English.

After a while I gave up, standing there wondering about our next move. Proceeding without permission would be rude. Returning another day would involve retracing the route to this point, which was long and tortuous. Moments later, the Navajo pointed behind him and explained in perfect English that the route we sought was right behind his hogan and that it was necessary to bear right at the clothes line, continue up a way and then go to the right even though the better road seemed to go left, because the left one was washed out not too far beyond the junction. Continuing along to the right would place us precisely where we wanted to be. Then he smiled. After providing instructions to the outcrop, he invited us to come back later to see his wife's fine rugs, for which she had won numerous awards. I thanked him, promised to return to see the rugs, and went on my way, puzzled by the delayed response but grateful for the information and the invitation back to the hogan. We found the outcrop, obtained our measurements and samples, and returned to the hogan to see what truly were magnificent Navajo rugs, several of which had large ribbons on them, awards from shows at the Museum of Northern Arizona.

I later learned that Navajos would never think of interrupting each other. They wait politely for the other person to finish, and watch for signs to indicate that the other person is indeed finished speaking before saying anything in response. In contrast, in typical Anglo fashion, we fill every slight pause by continuing to talk, dreading any moment of silence in a conversation. Without a clear sign that the other person is finished, a Navajo will wait patiently for the time when he can speak. It is a point that our own culture would do well to learn from the Navajo. In many ways, these are the same "listening" skills that help a geologist listen to the stories that the layers of rock have to tell. If we rush in with our hypotheses, we may be missing important information that the rocks will reveal to us, if we will only listen and observe patiently.

I learned a similar lesson from the Navajo people at the Monument Valley Tribal Park, which is owned and operated by the Navajo. We were camped at the Mitten Rock campground one night and had arrived too late to pay the fee at the entrance station. The sign indicated that a ranger would be around later to collect the fee should the station be closed. We started cooking dinner immediately, as it was late, and noticed just as we sat down to eat that a white pickup truck had stopped in front of our campsite. We assumed that we were expected to pay the fee, but the Navajo driver sat in his pickup without looking at us, and seemed to be focused on other business. Quite a while passed, and I thought perhaps we should offer to pay the fee since he did seem official and would probably want payment sometime during the evening. As I approached the vehicle, the young Navajo smiled broadly, clipboard in hand, and announced how much we owed for the campsite. It became clear that, indeed, he was there to collect the fee, but had patiently waited for us to make the first move. I paid him and returned to my dinner, wondering why he had been so patient and polite in the face of our seeming rudeness and indifference.

It became clear in reading one of Tony Hillerman's novels that it is considered impolite to approach someone's home uninvited. Navajo sit outside each other's hogans in their pickups and wait to see if someone comes out to welcome them in. Otherwise, it is assumed that company is not welcome at that time and the visitor then leaves, not wishing to intrude. The Navajo sense of place and respect for both the land and each other are deeply embedded in their culture.

At the end of field season each year, I leave the canyons and mesas of the Colorado Plateau to the more permanent dwellers of the land, and return to the office. The post-field-season blues set in. It is clear now, after many summers of fieldwork followed by many winters in the office, that the sense of the real, and the sense of connectedness that comes from being with the very Earth we attempt to understand, are elusive and fragile. It is difficult to maintain a sense of belonging to Earth when surrounded by concrete, cacophony, and computers. Joseph Campbell talked about the difficulty of holding onto the sense of awe and wonder that he felt in St. Patrick's Cathedral in New York City as he stepped out into the noisy traffic and bustling sidewalk along Fifth Avenue. For the field geologist, how can the feeling of the sacredness of the land be carried into a world of cyberspace?

5.4 THE FUTURE OF GEOLOGY

As valuable as geologic information is for addressing the pressing problems of our planet, it is possible that a field geologist's perspective, gained from an intimate connection to the Earth, is an even more crucial contribution to our modern world. A field geologist offers a perspective gained from studying the natural world in all its complexity, uncertainty, and wonder. Field geologists often take for granted our encompassing view of nature and time, and our ability to see into Earth. The probabilistic approach of geology and our appreciation for the interconnectedness of all things engenders an acceptance of the vagaries and uncertainties of nature.

The concept of ecological sustainability, much discussed of late, requires that we first establish the quality of life we seek to sustain. Even if we succeed in purifying our water supplies, removing pollutants from the air, finding renewable and clean energy sources, and mitigating the affects of natural disasters, we could still lead unsatisfying lives because we have become so detached from the very Earth we inhabit.

One wonders what the Anasazi who drank from the water hole at El Morro would think of the modern homeowner in Los Angeles, who hoses his sidewalks, unaware that entire rivers are diverted for him to have clean concrete. Societal detachment from the natural is pervasive, and our fascination with concrete and technology divorces us from a primordial relation to the land. To achieve sustainability, we need a sense of humility and awe for Earth. We also need an acceptance of the real and all that attends it—uncertainty, risk, wonder, connectedness, interdependence—an acceptance that is engendered only by encounters with the natural world.

I was heartened by a recent experience while lecturing about dinosaurs to a high school class in Denver. At the end of the lecture, the teacher asked me to describe for the students what it is like to be a field geologist. I started to describe the simplicity of a block of ice in the cooler, a full container of water, and the inexplicable joy of sleeping on the ground. A 16-year-old boy in the third row started to nod his head, and then his whole body, in agreement. His face broke into a broad Huckleberry Finn-like grin, in recognition of the simple life one experiences in nature. His resonance with the real gave me hope. There still exists a palpable longing for the simple things that are natural and the lessons Earth can teach us. Perhaps in a few years he will become a field assistant and learn the ways of field geology.

ACKNOWLEDGMENTS

For my many field colleagues, who know from whence I speak, and for the numerous wonderful field assistants who shared many glorious days in the field with me, I remain indebted. For the USGS and the National Park Service who supported my efforts to unravel the geologic stories of the Colorado Plateau, I am deeply grateful.

REFERENCES

Andrews, Sarah. Quote from a National Public Radio interview, May 1998.

Borgmann, Albert. 1993. *Crossing the Postmodern Divide*. Chicago: University of Chicago Press, 1992.

Chamberlain, T.C. 1897. "The method of multiple working hypotheses." *Journal of Geology, 5* (8): 837–848.

Covey, S.R. 1989. *The Seven Habits of Highly Effective People-Restoring the Character Ethic,* New York: Simon & Schuster: 150-156.

Wantland, Frank. 1998. "American Association of Petroleum Geologists explorer 'Toxic Worry Impairs Creativity'", 199(8):42.

PART II

THE EARTH SCIENCES IN THE LIFE OF THE COMMUNITY

6

A Multidisciplinary Approach to Managing and Resolving Environmental Conflicts

Brian Polkinghorn

Brian Polkinghorn earned an MS from George Mason University's Institute for Conflict Analysis and Resolution, and an MA and Ph.D. from Syracuse University's Maxwell School of Citizenship and Public Affairs, where he specialized in environmental conflict resolution. Polkinghorn has also been a National Environmental Management Program Fellow with the U.S. Environmental Protection Agency, where he analyzed the use of alternative dispute-resolution processes in regulatory and enforcement matters. Polkinghorn teaches in the Department of Dispute Resolution, Nova Southeastern University. His recent work includes assisting in the restoration of damaged coral reefs, resolving industrial siting disputes in Israel and the West Bank, and mediating disputes concerning utility reconstruction in Mostar and Sarajevo, Bosnia-Herzegovina.

In this essay Polkinghorn emphasizes the importance of a multidisciplinary or systems approach to environmental controversies. No one discipline holds the key to understanding or resolving our environmental problems. Polkinghorn also emphasizes that environmental decision making must be democratic rather than technocratic in orientation. The application of multiple disciplines should be balanced with the interests of nonexperts (e.g., local citizens or "stakeholders"). A systems approach, balanced with an appreciation of the role of the nonexpert, increases the possibility of avoiding conflicts.

Polkinghorn's account of the disciplines to be included within a systems approach consists of the physical, biological, and social sciences; the perspectives of the humanities are absent from this list. Polkinghorn sees questions of values as adequately treated through the social sciences. For when human values are understood to be subjective, there is no reasoning possible. The social scientist simply treats these claims as facts, neither right nor wrong, and tabulates the results.

"The Earth is one but the world is not." So opened the 1987 Brundtland Commission report, a major summary on the state of Earth and its many environmental problems. The statement summarizes the dilemma we face with many of our environmental issues. While the Earth system is an interrelated whole, humans comprise and control the social world—a realm fractured by countless conflicts over how we relate to and make use of the Earth.

This essay offers a model for overcoming these conflicts. It begins with the definition and clarification of concepts, which are used to construct a multidisciplinary model for managing environmental conflict. This is followed by an account of the set of tools gained through international field research and intervening in environmental conflicts. A case study is then presented that demonstrates the applicability and challenges of multidisciplinary environmental conflict management. The final section summarizes what has been learned to this point and issues a challenge to test the ideas that have been presented. Important lessons gained include the need to question more fully the various forms of expertise used in framing and discussing environmental problems, and the need to examine complex problems using a systems approach. Another point emphasized here is recognizing the inherently social construction of environmental problems and how worldviews, language, and human habits influence our relationship with the environment.

6.1 CLARIFYING CONCEPTS IN ENVIRONMENTAL DECISION MAKING

Our concept of the environment has evolved over the years and through various cultures to the point where almost any definition offered can be criticized for being too narrow, too broad, or somehow missing a key ideological point. One can think of the environment as being composed of everything experienced by sight, sound, smell, touch, or taste. However, this definition is unwieldy when it comes to regulating human behavior through policies, regulations, and management programs. One could also note that until relatively recently within Western culture, humans were largely considered to somehow lie outside the environmental context (see Everenden, 1985). Likewise, it has taken time for the collective consciousness of countries such as the United States to begin to more fully develop not only a philosophy about the relationship between humans and the environment but also to put it into place in legislative actions and environmental programs that acknowledge the environment (on these themes, see Bormann and Kellert, 1991).

We can examine the environment as a complex adaptive system containing interrelated elements or components that perform necessary processes. It is reasonable for instructional purposes to create a set of subcategories of environments, as long as it is understood that they are a set of abstractions useful for grasping a more complex and perhaps chaotic reality. One category is the *biological* environment, or biosphere, where life exists and where human impact is clearly evident. Marine biologists, zoologists, toxicologists, foresters, game wardens, botanists, and others study or work in such an environment. There is also the *physical* environment where the physicist, chemist, geologist, meteorologist, and others study and conduct research. Finally, there is the *social* environment where we live and work as well as transform the terrain to consume natural resources. This is where economists, political scientists, sociologists, anthropologists, and others study and work. None of these environments are mutually exclusive: As one system environment changes there is a tendency to alter the others. Figure 6.1 outlines the components that surround a systems approach to understanding the environment.

Thinking in terms of systems forces one to consider intentions and actions and their effect across the spectrum. It requires individuals to articulate clearly what they

A Systems Approach to Examining the Environment

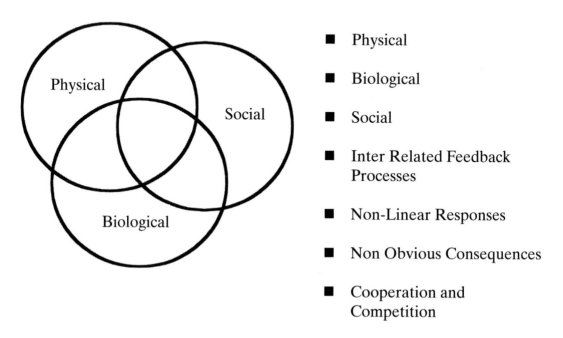

- Physical

- Biological

- Social

- Inter Related Feedback Processes

- Non-Linear Responses

- Non Obvious Consequences

- Cooperation and Competition

Figure 6.1   A Systems Approach to examining the environment.

think they know, what they may not know, and at least begin to consider what is truly unknown and possibly unknowable. More will be said on this later. For now, suffice it to say that decisions that are made without considering all these aspects can lead to unintended and unforeseen consequences, which in turn often lead to conflict.

Conflict is another one of those concepts we often use to describe many situations or events. Conflict arises when people are blocked from or denied access to the attainment of their goals. If one group is attempting to control or block a scarce resource such as freshwater, arable land, sources of food, free movement through a territory, or even less tangible things such as giving due recognition to another, then conflict can occur. Examples of environmental conflicts include the siting of a nuclear power plant or the poisoning of one's water supply by an upstream polluter.

A related concept is a *dispute*, which often focuses on a specific issue and the interests of the stakeholders (a stakeholder is anyone who has an interest or "stake" in a conflict or dispute). Disputes do not tend to consume as many of the stakeholders' resources because both the level of engagement and the importance of the issue do not typically cause people to go to such great lengths. An example of an environmental dispute is how to fine someone for illegally cutting down a protected tree. Disputes also do not typically impinge on such things as one's identity, nor do they tend to revolve around issues pertaining to the maintenance of basic human needs.

One problem with this definition of conflict is its anthropocentric or human-focused bias. For environmental conflicts also involve the needs of plants, animals, terrain, water, and other "things." The question arises, who represents plants and animals? The answer is that this is another source of conflict between humans, one that is often settled in courts and legislatures! The important factor to consider about conflict is that there are certain ways people think and engage one another (outlined in Figure 6.2) that may help make sense of complex disputes and conflicts.

Environmental problems, if not properly managed, can escalate into disputes and conflicts. In many instances environmental problems are framed and decided within a political context where people not only define the problem but often point to their opponents as the source. Most environmental conflicts are conceived, fought over, and decided almost exclusively within a social context with other system and environmental perspectives (physical and biological) acting as props or propaganda for a social agenda. It is common for people who claim to represent various aspects of the biological and physical environments to do so more for their own purposes than for the supposed needs of these environments.

This raises the question of how stakeholders tend to engage each other in environmental conflicts. Figure 6.2 depicts one way to analyze the sources of environmental conflicts. For now, note that as one moves from the top to the bottom of the triangle

Levels and Sources of Conflict

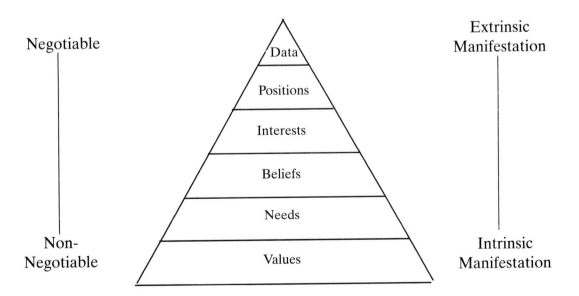

Figure 6.2   A Levels Approach to understanding environmental conflict.

the sources of conflict go from largely negotiable to nonnegotiable. Also note, that as one moves from top to bottom we shift from external and extrinsic matters to more intrinsic and personal areas. These will be discussed in detail below.

At the top of the triangle, in Figure 6.2, are the *data* that stakeholders use to explain how they see the event and to build their case for their proposed solution. When stakeholders disagree too much with each other's data, the discussion can deteriorate into a form of warfare over whose data are more accurate, legitimate, reliable, valid, and credible. Lying behind this discussion is often the erroneous belief that scientific data alone are sufficient to resolve environmental concerns.

At the next level of the triangle people will often resort to using data to express a strong opinion in the form of a *position*. This is easy to spot, because it is typically delivered as an ultimatum whereby parties take stands on what they are either willing or unwilling to do. Position-based discussions are typically adversarial, meant to force the other side to capitulate and are usually a quite ineffective means of trying to reach creative solutions. Fortunately, such deterioration of a discussion can be managed by reframing ideas in a way that moves people away from positional statements.

At the next level in the triangle are *interests*, and this is where most environmental dispute resolution (EDR) practitioners understand the idea that problem solving typically occurs. Interests tell us why stakeholders will or will not do something. Interest-based negotiations are designed to focus on the "who, what, when, where, and why" questions that stakeholders can answer. Interests allow flexibility, whereas positions are static and unambiguously one-dimensional demands. We will return later to the focus on issues when we examine our case study.

As we move down the triangle the area we next encounter becomes more focused on individual personal *beliefs*—often a difficult place to start a negotiation or to make decisions. One's belief system is a set of ideas and rules regulating how we act and think and can be a highly personal and sensitive topic of inquiry. Inquiries into this area are thus often perceived as a challenge to the way one has constructed a belief system and understanding of the world, which is tantamount to questioning their credibility, personal integrity, and intelligence. A typical response is to become defensive, even though it is possible that these beliefs are indeed the source of conflict.

There are generally three ways of approaching an environmental conflict that has been framed in the parties' beliefs. The first is to do nothing and let it run its course. The theory goes that there are some conflicts that need to be allowed to go forward because of the potential to produce more benefit that harm (Kriesberg, 1998). The second approach is to engage in a winner-take-all battle over whose beliefs are paramount. For instance, one can believe in conservation measures and be opposed to the construction of housing in the breeding ground of a migratory bird. A typical response might be to have the area declared a bird sanctuary, which transforms one's beliefs into action. However, this is at the expense of other parties' beliefs. A third, more difficult but hopeful approach is to try to translate one's beliefs into some interest-based discussion. The real question is, what are the parties willing to do to protect this area and yet allow other interests in development to be addressed? Conflicts do not always need to be a win–lose proposition.

When conflicts begin to focus on *values* and basic human *needs* the problem often becomes so difficult, intense, and stressful that it is nonnegotiable. This is because needs and values are fundamental to physical and mental existence. Violations of basic principles or values, threats to one's safety, security, possessions, territory or identity may produce this type of response, even leading to violence. Much of the literature in environmental dispute resolution cautions against intervening in value conflicts, although some have gone so far as to say that there are some justifiable reasons and relevant information that can result in this type of endeavor (Susskind and Field, 1996).

Figure 6.2 also provides a way for us to analyze how people think about and direct a conversation, be it a discussion, debate, or heated conflict. By understanding how data, position, interests, beliefs, needs, and values can be expressed concerning the same topic, we become aware of ways on how to manage heated debates and keep an impasse from occurring. This raises the question of the source of people's thinking. To what degree does the way a person is socialized or professionally trained have an influence on how the person thinks? How does level of education, occupation, or personal traits such as race, ethnicity, political orientation, and gender influence how people will frame and discuss an environmental conflict? Is one approach to framing an environmental problem necessarily better than another? Is one discipline somehow better at finding the most optimal solutions to many of our environmental problems? These questions lead to the topic of single discipline and multiple discipline thinking.

6.2 UNDERSTANDING SCIENTIFIC PARADIGMS AND PROBLEM SOLVING

Nature is not as clearly understood as we would like to believe; as a result, the way environmental problems are framed and debated can become a basic source of conflict. Recent research on human-induced environmental problems indicates that we can expect such conflicts to increase over the next generation (Homer-Dixon, 1999.) For instance, when an environmental dilemma arises over how to regulate underground storage tanks, there are politicians, policymakers, regulatory agencies, citizen groups, environmentalists, scientists, labor and business organizations, and various other stakeholders who express their own opinions and ideas on how to solve the problem. Their opinions, ideas, expertise, vocabularies, and worldviews can become a source of conflict, as people use language and perspectives unfamiliar to one another. The way we communicate about environmental problems is also a source of conflicts (for instance, see Williams and Matheny, 1995).

One point worth noting: When environmental problems arise, people have a tendency to rely on their respective form of expertise to address it and offer solutions (for an excellent example of this phenomenon, see Sapp, 1999). This highlights another, less obvious source of conflict—namely that not all forms of expertise can adequately address a particular environmental problem. As a result, the often unstated debate involves claims that one type of knowledge and expertise is more pertinent and important than others.

Single-discipline approaches to environmental decision making often have a built-in problem, which can be described by making generous use of Thomas Kuhn's (1970) notion of "paradigm." A paradigm is a collection of concepts, theories, and overall worldviews that dominate the way a school of thought or an academic discipline frames and solves problems. Each discipline relies upon its own disciplinary paradigm to frame an environmental problem. Yet if we get caught in a debate as to which concepts or theories are the most "correct" for framing a problem, instead of creating a multiple disciplinary means of decision making, we lend credibility to the idea that one form of knowledge and expertise is paramount. A useful analogy is that if you are a carpenter and the only tool you have in your toolbox is a hammer, eventually all your problems begin to look like nails.

Kuhn provides examples from physics that demonstrate that many groundbreaking discoveries come from either younger or nonindoctrinated scientists from outside the field. What these people have in common is an approach that is based on either altered or different concepts, theories and ways of framing problems, or a healthy skepticism concerning how the discipline is framing problems in the first place. The important point is that these people are capable of working outside the normal problem-solving format to find solutions to perplexing problems that their colleagues were incapable of discovering.

How does this fit into a model of environmental management? First, it provides a conceptual idea, namely paradigms, that helps us understand the variety of ways that people approach the study of the environment. For the sake of ease, a continuum is offered here to describe the variety of approaches that various disciplines use to study and manage the environment. It is important to realize, though, that science paradigms are only a few of the ways we can examine environmental problems. Basic lived experience is another way of understanding how humans interact with the environment (see Table 6.1).

Table 6.1 immediately exposes a number of differences between the sciences. For a discipline such as physics may establish facts that can be so generalizable over time and place that they become laws. This form of empirical truth is less applicable when the focus shifts to disciplines such as ecology or the social sciences. At best, in ecology,

Table 6.1 CHARACTERISTICS OF DISCIPLINES AND ENVIRONMENTAL DECISION MAKING

	Physics Chemistry *"Hard Science"*	Ecology Biology *"Mixed Sciences"*	Sociology Anthropology *"Soft Sciences"*
General Focus	Deterministic Positivistic Empirical "Truth"	Mixed science methods	Experiential Exploratory Empirical "Fact"
Types of Logic	Largely deductive	Deductive—Inductive	Largely inductive and some deductive
Typical Research Methods	Basic and applied theoretical and scientific control	Typically applied controlled and field studies	Mostly applied Mixed methods Field studies
Facts	"Natural" Reflects true quantities and qualities	"Natural" and institutional (descriptive)	Institutional Descriptive and prescriptive
Theory	Focus on both theoretical and concrete Deterministic	Partially deterministic	Limited reliability, validity, and generalizability (No Grand Theory)
Forms of Uncertainty	Errors in experimental method	Error partially due to design or change in the environment	Error-prone on many counts; difficult to establish causation

models can be constructed to predict how systems operate or how people will behave under specifically controlled situations. In contrast, social sciences are inherently prone to problems concerning reliability, generalizability, and the effects of the subject's awareness of being part of an experiment or study.

Disciplines also tend to have an established sense of logic. In theoretical physics one can see deductive theory building being applied to research agendas, while in other disciplines, where the level of empirical truth or certainty is much less, we see researchers examining environments for clues in a more inductive manner. Fieldwork in geology or biology is an inherently inductive process. The approach to making sense of bits and pieces of data can also be found in many of the social sciences. However, both deductive and inductive inferences are a part of the paradigms that make up many disciplines. The important point is to emphasize the different reasoning approaches in the various disciplines, and how each brings strengths and weaknesses to an environmental problem.

Facts are not that easy to describe, much less find. Certain laws can produce facts that are indisputable. Many "facts" are subject to interpretation, which again is influenced by how people are trained to think. Many "facts" are actually causal descriptions of how we believe things work. For instance, if we live next to a creek that has been poisoned by PCBs we may also extrapolate from that fact to the assumption that aquatic life in the creek is also being influenced. A toxicologist can employ training and research methods to provide us with a greater understanding of the situation and by clarifying facts. The important thing to keep in mind is that various aspects of the environment possess different types of facts, which offer different and even irreconcilable views to how the environment operates.

When it comes to understanding how theories influence environmental decision making, perhaps the most hotly discussed issue revolves around levels of certainty. At one end of the continuum are deterministic theories that claim to explain not only how things happen, but also what you can expect to occur regardless of the time or place in which it happens. At the other end of the continuum are those theories that are constructed by observation, exploration, and discovery. Theory development can either be a strictly cognitive exercise, or it can occur through prolonged observation of events such as animal behavior and seasonal variations in plant life.

Paradigms have a way of clouding the degree and form of uncertainty that occurs. The level of uncertainty and types of assumptions that are made in calculating relationships can be rather astounding. How, for instance, can we know how much water there presently is in the Everglades? How much of it is brackish? How many alligators are there? Where are they? Do water conditions influence alligator mating and the success of live offspring? Many of the answers are based on estimates that can easily vary by an order of magnitude or more.

Finally, what is usually missing in any account of the utility of various disciplines to solve environmental problems is an examination of what it means to be an expert. The taxonomy just presented should not be seen as an elitist model whereby only people with a given academic knowledge are allowed to solve environmental problems. As stated at the beginning of this part of the discussion, it is unwise to rely strictly on science in solving such complex problems.

Everyday people can be experts on some aspect of the environment. This untapped resource is important in understanding how conflicts arise and who is needed to help solve them. For instance, in the United States the environmental justice movement focuses on the disproportionate impact of environmental hazards upon certain segments of the community. One of the demands of this movement is that we put people back into the environmental problem-solving equation. Works by Bullard (1990, 1993), Byrant and Mohai (1989), and Hofrichter (1993) point to the unequal treatment of some groups when it comes to sharing the costs and benefits of modern society. In essence, their argument is that race and class are the determining factors in where locally unwanted land uses (LULUs) such as landfills will be located. People living near these LULUs have brought to light another meaning of expertise. In one recent case a participant shouted, "I am the expert of those who live in a toxic dump!" This exposed the entire group not only to the way we were labeling each other as certain types of experts, but also that being an expert in some subjects does not require an advanced degree. More importantly, this event also raised the notion that it is possible to be an expert in environmental problems that may involve a combination of various disciplines as well as an accumulated set of life experiences.

The environmental justice (EJ) movement has clearly taken on this debate about who makes decisions concerning the protection of human health and the environment, and is increasingly taking part in many major environmental decisions. The EJ movement has had its critics. They have been accused of being NIMBYs or "Not in my back yard"; CAVE (citizens) or "Citizens against virtually everything"; and BANANAs— those who believe we should "build absolutely nothing anywhere near anything." But in essence, EJ represents people who have become invisible within political decision making and problem-solving processes. It represents a part of environmental decision making where expertise is not a function of what you know, but who you are.

In some instances, environmental problems can be addressed using one disciplinary school of thought. However, many environmental conflicts require the expertise of numerous forms of knowledge, and therefore some coordinated system or model will be necessary to reflect existing conditions accurately. This approach has been given many names, including "systems theory" and "holistic management."

In my experience, there are a number of ways in which environmental problems and decisions are framed. The first is legislative and legal in orientation. In some instances, Congress makes clear the goal of a policy or regulation. In other cases, Congress creates wide-sweeping and purposively vague legislation that allows policymakers to judge for themselves the best way to carry out the congressional mandate. Here management strategies originate through the discretion of federal or state agencies. In still other instances, the ways an environmental problem are managed comes from real and imagined emergencies. In the United States we have experienced rapid environmental decision making over things such as asbestos and alar. In both instances it was public awareness and public concern that started the environmental management machinery. However, both cases highlight the need to create a process that is framed not so much in reaction time but on established criteria that focuses on first framing a problem so to address it in the most effective manner.

With these points in mind, we can move to a case study that does precisely this, using a multidisciplinary approach that takes stakeholder interests seriously.

6.3 CASE STUDY: EXPANDING ROUTE 1 INTO THE FLORIDA KEYS

In 1997, numerous state and federal agencies engaged residents of the Florida Keys as well as Monroe, Miami-Dade, Broward, and Palm Beach counties in a series of public dialogues on a proposal to expand Route 1 between Florida City and Key Largo, Florida. Florida City is the last major municipality before taking the "18-mile stretch" of a mostly one-lane road that connects to Key Largo, the first or northernmost Key. This is the major route into and out of the Florida Keys. From Key Largo, Route 1 continues roughly another 120 miles to its terminal point in Key West, the southern-most Key.

This section of road is mostly single lane in each direction, with two staggered short stretches for the north and southbound traffic to utilize an independent passing lane. As a result of this design, the 18-mile stretch is a dangerous section of Route 1 with frequent and horrific head-on collisions. To compound the problem, the entire section of road is within a wilderness setting. The difficulty in responding to the scene of an accident contributes to the high number of fatalities. The general question posed to the public was "Should this 18-mile stretch of road traversing the southeast portion of the Florida Everglades and Florida Bay be expanded to accommodate more traffic, and if so what should the plan look like?"

Another important reason for expanding the 18-mile stretch was that in the event of a hurricane, the state of Florida wanted to decrease the evacuation time of the entire Florida Keys. The road design has been blamed for numerous and frequent acci-dents that sometime closed the road, and this impacts evacuation time. Also, if south-bound traffic is allowed to flow as a hurricane is approaching, this limits the evacuation to one lane. In addition, authorities typically wait for an official warning to be declared before closing inbound traffic, meaning there are only 24 hours to evacuate people from the Keys. If an accident were to occur, then the access route could be slowed or even closed, leaving people exposed at sea level to the oncoming force of a hurricane.

As evidence of this concern, South Florida public safety officials in the Division of Emergency Management recalled their growing alarm as Hurricane Andrew came bearing down on South Florida in August 1992. They decided that if the path of the hurricane were to move even slightly farther south and into the Keys they would then call the Federal Emergency Management Agency and request an additional 6,000 bodybags to accommodate the anticipated number of victims. Fortunately, Hurricane Andrew did not veer south into the Keys. This explanation for the road expansion, at first and by most accounts, seemed to be based on a reasonable need to be prepared in the event of a hurricane.

In preparation for public dialogue, a number of plans were constructed with the help of engineers on a variety of roadway designs. These plans and additional informa-tion were meant to enhance the discussions and have information readily available for public review. Some ideas included a one-lane setup with a middle passing lane run-ning the entire 18-mile length of the highway. Another had a two lane system and yet another incorporated a two-lane system with another middle passing lane. In addition, much attention was also placed on elevating a bridge so as not to disrupt traffic flow on this stretch of Route 1. The most ambitious plan was to elevate the entire 18-mile road-way so as not to interfere with the southward-moving flow of surface water that is vital to the Keys' ecosystem. All these plans, ideas, and good intentions gave many people the feeling that the road expansion was essentially a done deal. This in turn led to the heightening of opposition.

Perhaps the most interesting opposition to the proposed road expansion came from groups of people living in the Keys. Overall, many residents of the Keys have a unique view of the world, and maintain a decidedly different culture from the "mainland" folk. For instance, back in the early 1980s, citizens living in the Keys (an area known to locals as the Conch Republic) declared war on the United States government, immediately surrendered, and sought reconstruction loans. This ploy to shore up a sagging economy is both amusing and half worrisome: Some of the sentiment behind this stunt was rather heated when it came to federal agencies such as the National Oceanographic and Atmospheric Administration (NOAA), who many saw as the primary cause of financial suffering in the Keys. NOAA is in charge of marine sanctuaries, the protection of grassy beds and, most importantly, coral reefs. All these are places where locals have made their living for generations and are now being forced to contend with severe NOAA restrictions. Opposition to the road expansion was expressed by many residents as a form of invasion, another attack on their culture, and, furthermore, a way to flood the Keys with too many people who would drive up property values to the point where locals would be gentrified out of their own homes. If anything, residents of the Keys generally felt it would be best to leave the road alone.

Each group that had a stake in the road expansion issue had an agenda and often expressed it in full detail. Soon there were many groups officially involved, adding considerable complexity to the discussion of possible road expansion. Table 6.2 is a partial list of stakeholders. The sheer number emphasizes just how quickly a supposedly simple policy question can become complex and confusing.

At some point prior to public involvement a series of decisions on the part of the state of Florida and the Army Corps of Engineers were made on how to collect information, provide relevant avenues for public input, and what to do with the information. The intriguing point of this case is that instead of the usual public-meeting format where people expressed themselves in a series of monologues to government decision makers, various meetings were convened throughout South Florida at places such as Nova Southeastern University, where the interaction was deliberately placed within a small-group format. People from different constituencies worked together on defining what they thought the road expansion entailed as well as some of its possible consequences. As a result, a number of interesting points emerged that may not have been identified in traditional public forums.

For instance, these are just a few of the issues that were identified:

- New road construction would require more elaborate runoff and collection pits.
- A raised bridge would alleviate problems with sheetflow of water (across road) but would require an even more elaborate runoff collection system.
- Increased access to the Keys could or would:
    drive up property values;
    alter the generations-old culture of the local communities;
    increase development and pollution;

Table 6.2 STAKEHOLDERS PARTICIPATING IN THE ROUTE 1 EXPANSION MEETINGS

Federal and State Groups

U.S. Army Corps of Engineers	Florida Keys National Marine Sanctuary
U.S. Environmental Protection Agency	Governor of Florida
U.S. Department of Interior	Florida Department of Transportation
Federal Highway Administration	South Florida Water Management District
U.S. Department of the Navy	Florida Department of Natural Resources
National Oceanographic and Atmospheric Administration	Florida Department of Health and Rehabilitative Services
Everglades National Park	Florida Highway Patrol
U.S. Congress	The Game and Fresh Water Fish Commission
Florida Department of Community Affairs	Real Estate Division of Construction and Land Management Department
Florida Department of Emergency Services	

Local Governments and Groups in the Florida Keys

Monroe County Board of Commissioners	Chamber of Commerce for the Florida Keys
Monroe County Land Authority	Chamber of Commerce for Key Largo
Monroe County Emergency Management	Chamber of Commerce for Homestead and Florida City
Monroe County Tourist Development Council	
Monroe County Planning and Zoning	Florida Keys Motel Brokers Inc.
Monroe County School Board	Florida Marine Research Institute
Monroe County Sheriffs Office	Center for Marine Conservation
Monroe County Growth Management and Public Works Division	

National and Local Nongovernmental Organization Groups

American Red Cross	The Coral Reef Coalition
Greenpeace	The Florida Keys Citizen Coalition
The Nature Conservancy	The Islamorada Charter Boat Association
The Wilderness Society	The Organized Fisherman of Florida
Audubon Society	The Florida Land and Sea Trust
Botanical Society	The 1,000 Friends of Florida
	The Florida Bay Initiative Inc.
	The Florida Keys Fishing Guide

---

overburden already short water supplies and drive up water prices;

produce too much sewage and pollute Florida Bay, the Atlantic Ocean, and the upper Keys;

increase sewage and other pollution that could damage coral reefs (recent research had made the link between migrating sewage and reef damage)

- Leaving the Keys during a hurricane is a voluntary decision. People cannot be forced from their homes, so the evacuation explanation for expanding the road leaves one to wonder about the efficacy of the plan.

Retaining a systems approach to framing environmental (and in this case developmental) problems allowed participants and decision makers to see how an option that focuses on one system can influence another. Also, instead of adopting a single frame of reference to make decisions on whether or not to undertake the road expansion, it was recognized that any decision should involve input from many professionals. Some of the experts came from civil engineering, marine biology, hydrology, geology, toxicology, economics, and researchers in environmental justice. In essence, the substance of the road expansion required a multidisciplinary approach with the full knowledge that the decision would be framed in a political setting.

In terms of understanding the sources of potential conflict, the systems approach effectively limited debates over data and the legitimacy issues that often arise. Trained third-party facilitators were also attentive to the hard positions that some of the participants held and helped them articulate their frustration, anger, anxiety, and fears. To get issues out and ideas and options generated, a deliberate attempt was made to hold off discussing specific and detailed data. Also, an early focus on the interests of the stakeholders helped in getting more information out early for others to hear. Whenever people began to engage others in more position-based arguments, attempts were made to get people to think why they were insisting on such solutions. In essence, the focus was shifted off people and back onto the problem, with an emphasis on what interests were being promoted or curtailed by the proposed expansion of Route 1.

In the end, the U.S. Army Corps of Engineers in Jacksonville, which has control of the road, the intracoastal waterway, and the entire water distribution system within the Everglades region, decided against expanding the road base. The rationale for widening the road so as to increase evacuation efforts in the Keys was not seen as being so persuasive as to offset the potential negative consequences of expansion. In fact, the original stated intention to enlarge the road to decrease evacuation time in a hurricane was tested in September 1998 when Hurricane Georges came through the Keys, with the eye of the storm crossing directly over Key West. Although there was extensive damage throughout the Keys, the evacuation ran smoothly and few people died.

In the spring of 1999, however, a new proposal was made for a three-year multimillion dollar study on expanding the 18-mile stretch of Route 1. Thus, the debate continues. But the way it was managed the first time is, according to the stakeholders involved, a powerful model to follow.

6.4 SYNTHESIZING IDEAS

"The earth is one but the world is not." The meaning of this phrase may never be agreed upon, but it certainly does encourage one to think about how environmental problems arise and escalate into heated disputes—or worse, into full-scale conflicts where people become stubborn or even violent. Throughout this essay, the idea that Earth is one and the world is not has been the underlying assumption behind the way individuals and groups examine human relationships with their environment. We do not always agree on what constitutes the basis for an environmental problem. Nor do we always agree on how to frame a problem, much less how to approach it and search for solutions.

This essay has offered several points to consider when examining environmental problems. The first is to view the environment from a more holistic reference point. In terms of deciphering how elements or components in an environment interrelate—an incredibly complex task—the suggestion put forth here is to adopt a systems approach. Learning what constitutes a system and how components of a system relate to each other is a challenging task. The ways in which local events may be impacted by global change exemplify both the importance and difficulties confronting a systems approach. In fact, biologists and ecologists often struggle over the notion that we can clearly separate one system from another, which challenges our overall conception of Earth and environmental systems as well as the assumption that systems perform a "boundary-maintenance" function.

Second, we must recognize the limitations of our intellectual capabilities. There are aspects to Earth and its dynamics that we simply don't understand, and are unlikely to any time soon. Given this uncertainty, we must resist the urge to "win" when it comes to framing, examining, debating, and contemplating complex and often poorly understood problems between humans concerning their environment. And third, we need to focus on the protection of Planet Earth in order to create the groundwork for effective management of environmental conflicts. Perhaps this challenge can also alter the phrase at the beginning of the essay so that the world, if not one, can at least engage in one common conversation.

REFERENCES

Amy, Douglas J. (1987). *The Politics of Environmental Mediation*. New York: Columbia University Press.

Bormann, F. Herbert, and Stephen R. Kellert (eds.). (1991). *Ecology, Economics, Ethics: The Broken Circle*. New Haven, CT: Yale University Press.

United Nations World Commission on Environment and Development. (1987). *Our Common Future*. "Brundtland Commission Report." New York: United Nations.

Bryant, Bunyan, and Paul Mohai (eds.). (1989). *Race and the Incidence of Environmental Hazards: A Time for Discourse*. Boulder, CO: Westview Press.

Bullard, Robert. (1990). *Dumping in Dixie: Race, Class and Environmental Quality*. San Francisco: Westview Press.

Bullard, Robert. (1993). *Confronting Environmental Racism: Voices From the Grassroots*. Boston: South End Press.

Burton, John (ed.). (1990). *Conflict: Human Needs Theory*. London: Macmillan.

Everenden, Neil. (1985). *The Natural Alien: Humankind and Environment*. Toronto: University of Toronto Press.

Hofrichter, Richard (ed.). (1993). *Toxic Struggles: The Theory and Practice of Environmental Justice*. Philadelphia: New Society Publishers.

Homer-Dixon, Thomas F. (1999). *Environment, Scarcity, and Violence*. Princeton, NJ: Princeton University Press.

Kriesberg, Louis. (1998). *Constructive Conflicts: From Escalation to Resolution*. New York: Rowman and Littlefield.

Kuhn, Thomas S. (1970). *The Structure of Scientific Revolutions* (2nd ed.). Chicago: University of Chicago Press.

Lemons, John (ed.). (1996). *Scientific Uncertainty and Environmental Problem Solving*. Cambridge, MA: Blackwell Science.

Lee, Kai N. (1993). *Compass and Gyroscope: Integrating Science and Politics for the Environment*. Washington, DC: Island Press.

Lein, James B. (1997). *Environmental Decision Making: An Information Technology Approach*. Malden, MA: Blackwell Science.

Sapp, Jan. (1999). *What Is Natural? Coral Reef Crisis*. New York: Oxford University Press.

Savory, Allan, and Jody Butterfield. (1999). *Holistic Management: A New Framework for Decision Making*. Washington, DC: Island Press.

Susskind, Lawrence, and Patrick Field. (1996). *Dealing With an Angry Public: The Mutual Gains Approach to Resolving Disputes*. New York: The Free Press.

Williams, Bruce A., and Albert R. Matheny. (1995). *Democracy, Dialogue, and Environmental Disputes: The Contested Language of Social Regulation*. New Haven, CT: Yale University Press.

7

Science and Environmental Policy: An Excess of Objectivity

Daniel Sarewitz

Daniel Sarewitz is senior research scholar at Columbia University's Center for Science, Policy, and Outcomes, and author of Frontiers of Illusion: Science, Technology, and the Politics of Progress *(Temple University Press, 1996). He received his Ph.D. in the Geological Sciences from Cornell University in 1985. At the same time, Sarewitz reports that he noticed a widening chasm between the technical knowledge that he was acquiring, and his personal experience in the surrounding world.*

In 1989 Sarewitz left science to work for the U.S. Congress—an institution that he describes as being in many ways the antithesis of academe. Congress made clear to him what science had sought to obfuscate or even deny: that reality operates on many levels simultaneously, that these multitudinous realities are often conflicting and incommensurable, and that there is no possible long-term equilibrium state in which such tensions are resolvable. For four years, Sarewitz worked for Congressman George E. Brown, Jr., Chairman of the House of Representatives Committee on Science, Space, and Technology.

In this essay, Sarewitz looks at the capacity of science to help resolve environmental conflicts. Scientists and decision makers alike tend to view the role of science in environmental policy as prescriptive. The goal is to create objective information that can cut through the morass of politics and enable wise decisions. Sarewitz claims that in the real world this happy result rarely emerges. What one finds instead are politicians using science to back their political positions; or even arguing over the technical merits of the science, rather than about the societal merits of the politics. But rather than seeing this as a problem caused by politicians distorting the scientific facts for partisan purposes, Sarewitz suggests another possibility: that nature itself resists unitary characterization. The appeal to science to resolve our environmental questions thus presents us with an "excess of objectivity."

In the mid-1980s, a nasty academic conflict flared up over the very existence of objective, scientific knowledge. Later dubbed the "science wars," this conflict apparently pitted the natural sciences against the social sciences in a debate not just about the process of research, but also the nature of scientific facts. As in many academic battles, genuine intellectual discourse was quickly overwhelmed by a rising tide of rhetoric. In the heat of battle, all nuance was lost in the quest for victory, and a single, black-and-white question came to dominate the contest: Does science achieve an objective view of nature, or are all scientific facts constructed by social interactions? The latter, "constructivist" view considers "the 'truth' or 'falsity' of scientific claims…as deriving from the interpretations, actions, and practices of scientists rather than as residing in nature."1 More contentiously still, "the settlement of a [scientific] controversy is the cause of Nature's representation, not its consequence."2 Like red capes before bulls, such pronouncements drove some natural scientists to rage: "[The] logic of cultural constructivists seems to us sloppy and full of holes…their evidence dubious, and their case corrupted by special pleading and covert appeal to emotion."3

As for the public debate, the natural scientists, not suprisingly, soon had the constructivist social scientists on the run. Mostly famously, the physicist Alan Sokal managed to perpetrate a humiliating hoax on social scientists when his patently nonsensical constructivist critique of quantum physics was published by the non–peer-reviewed journal *Social Text*.4 This triumph was duly debated in the pages of the *New York Times*.5 Other "victories" included the resignation of *Science* magazine's book review editor, reportedly in response to criticism she received after printing a negative review6 of *The Flight From Science and Reason*,7 a collection of anti-constructivist polemics; and the decision by the Institute for Advanced Study at Princeton not to offer a job to a highly regarded historian of science with a known sympathy for the constructivist position, even though he had been approved by a hiring panel. The constructivists were no match for the institutions of modern natural science; invoking imagery of the Spanish Inquisition, *The Economist* wryly observed: "You hear at times the sound of butterflies being broken in cyclotrons."8

Lost in the din of battle was the possibility that both sides were right. In the real world, the success and impact of science is argument enough for the validity of its method and results, socially constructed as they may be. As David Hull writes: "No amount of debunking can detract from the fact that scientists do precisely what they claim to do."9 Nor can it be denied that science generates reliable knowledge that can be used, for example, in the design of technology. Yet this observation can and does coexist side-by-side with another reality: Society and culture create a context within which knowledge is pursued and used, and they influence both the types of facts, and portrayal of the facts, that we acquire.10 The history of geology after World War II illustrates this observation. Would there have been a plate tectonics revolution if the Cold War had not subsidized the seafloor mapping and global seismic arrays that led to the recognition of mid-ocean spreading and plate subduction?

Facts are both objective (that is, representations of something real) and constructed (that is, products of social context). In this essay, I will discuss the interaction of the objective and the constructed (without, I hope, resort to more social science jargon) in the arenas of politics in general, and environmental policy in particular. This interaction is of growing importance in a world where the character of environmental problems becomes ever more global, more severe, and more divisive—and where science is increasingly called upon to mediate and solve these problems.

A fundamental observation is that a desired goal of science in environmental policy—to help provide answers that can resolve political controversies—can rarely if ever be achieved. I will argue that this goal is illusory not because science fails to contribute objective facts to our arsenal of knowledge, but because it does so all too well. Science is so effective at generating facts that we are saturated with objectivity, to the point that, in the political world, science often does us very little good at all, and sometimes makes considerable mischief.

7.1 SCIENCE TO THE RESCUE?

In developing this argument, we must first establish that science is called upon to help society resolve difficult and contentious environmental problems. And in fact, the federal government spends billions of dollars on research aimed at solving or clarifying or providing guidance on environmental or natural resource controversies. The $1.8 billion per year U. S. Global Climate Change Research Program (USGCCRP) is only the largest and most conspicuous of these efforts. And the expenditures for this program are explicitly justified in terms of their value for making policy:

> The U.S. Global Change Research Program was conceived and developed to be policy-relevant and, hence, to support the needs of the United States and other nations by addressing significant uncertainties in knowledge concerning natural and human-induced changes in the Earth's environment....The USGCCRP is designed to produce a predictive understanding of the Earth system to support national and international policymaking activities across a broad spectrum of environmental issues.11
> A better understanding of the science of climate change is critical to determining the appropriate global mitigation and adaptation policy.12

Similar rationales underlie or have underlain research programs on acid rain, nuclear waste disposal, oil and gas reserve estimates, endangered species, air quality, and a host of other environmental and natural resource controversies. Central to these rationales is the idea that by introducing science, and the objective information that science can produce, into an environmental controversy, rational policy solutions will be facilitated. This idea is illustrated in Figure 7.1, which depicts how the process of integrating science into environmental policy supposedly works. As suggested in this flowchart, the process is linear and progressive, starting with the identification of a problem and proceeding in an orderly fashion through scientific research and predictive modeling, at which point the science is introduced into the political process, policies are developed, and solutions are reached.

Central to this scenario is the apparently self-evident idea that scientific research can provide the basis for political action. This notion is deeply embedded both in the science and the policy communities, to the extent that when a new environmental controversy begins to emerge—sometimes as a result of scientific research—the instinctive reaction is to call for more research. Such a seemingly involuntary response was on display recently when the U.S. Environmental Protection Agency (EPA) tried to promulgate new regulations governing the emission of very fine particulate air pollution. Congress reacted in an entirely predictable fashion: Those who were opposed to the regulations called for more research.[13] Of course, such a reaction is, to some extent, a tactic aimed at delaying implementation of environmental regulations without appearing to be obstructionist, but it also reflects a strong faith that, at some point in the

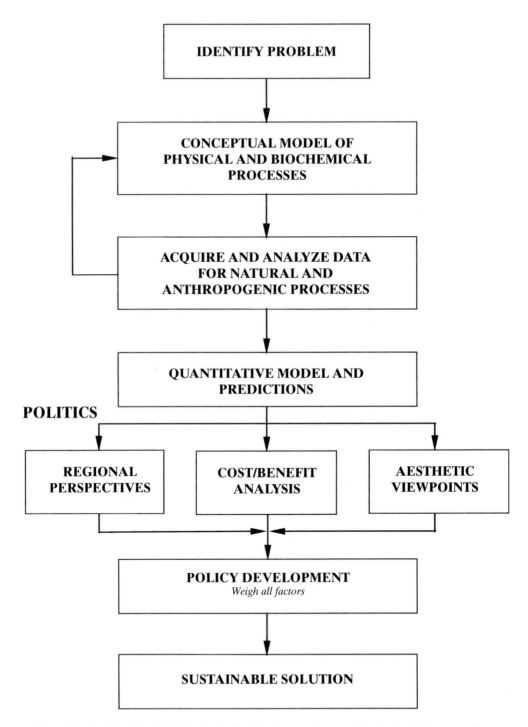

Figure 7.1   Traditional model ("physics view") of the linear and sequential relationship between science and politics in environmental decision making.

future, sufficient new knowledge will be acquired to stimulate the necessary action to resolve the problem "rationally."

This mental model of how science can contribute to environmental policy-making is consistent with the norms of a culture that places great faith in science and the rationality that science can deliver. Yet, in the real world, our expectations for science in policy-making are often confounded. Rather than resolving political debate, science often becomes ammunition in partisan squabbling, mobilized selectively by contending sides to bolster their positions. Because science is highly valued as a source of reliable information, disputants look to science to help legitimate their interests. In such cases, the scientific experts on each side of the controversy effectively cancel each other out, and the more powerful political or economic interests prevail, just as they would have without the science. This scenario has played out in almost every environmental controversy of the past 25 years. Even when science is alleged to have played a decisive role in resolving a policy dispute, as in the case of the international ban on production of chemicals that deplete stratospheric ozone, a closer look at the politics usually shows not that the science convinced policymakers to take the correct action, but that the science and the prevailing political interests fortuitously converged.[14] Indeed, in the area of global warming, where a highly touted (but, as will be shown, illusory) "consensus" of experts has publicly warned of the need to take international action to curtail greenhouse gas emissions, powerful opposing interests in the United States have ensured that no meaningful action has been taken.

However, my point is not that money or vested interest wins out over science and "rationality" every time. The plethora of environmental laws and regulations that were implemented in the United States from the late 1960s through the 1970s, despite the often energetic opposition of private industry, shows that widespread popular support for environmental action can defeat powerful monied interests. But it would be a mistake to suggest that this support was underlain or created by a strong and irrefutable scientific base, and that environmental policies adopted by the government were always dictated by science. Rather, these policies were a response to a wave of popular support that reflected evolving aesthetic, ideological, and ethical perspectives about the preservation of nature and the protection of environmental assets such as clean water, clean air, and endangered species. The science that was used to support such perspectives was often suggestive but rarely if ever uncontestable. At the same time, the growing public commitment to environmental protection was itself a major stimulus of new scientific research. That is, the politics behind environmentalism was probably more important for furthering the science than the science was for advancing the politics.

Nor is it productive to blame politicians for manipulating or distorting objective science to support partisan positions. Naturally, politicians will look for any information or argument they can find to advance their agendas—that is their job. While politicians may not be above playing loose with scientific truth, more often they can and will simply search out—and find—a legitimate expert or two who can marshal a technical argument sympathetic to the desired political outcome. It is the job of politicians to play politics, and this—like the second law of thermodynamics—is not something to be regretted, but something to be lived with. Given this reality, while scientists and politicians will often, as a matter of course, prescribe "more research" to resolve a political dispute about the environment, their prescription rarely has its desired effect over the relatively short time scales within which political decisions must be made—from a few weeks to a few years—and may in fact make things worse by providing more fuel to enflame debate.

7.2 THREE OBSTACLES TO BEING RESCUED BY SCIENCE

The idea that facts provided by science can help resolve political controversies related to the environment falters on three obstacles at the interface of politics and science. First, the goals of politics and science are different and often contradictory. Second, politics expects predictive certainty from science, but the complexity of nature invariably confounds this expectation. Third, the scientific view of nature is sufficiently rich and diverse to support a diversity of strongly held and often conflicting political interests and public values.

THE GOALS OF POLITICS AND SCIENCE

In a democracy, the political process is aimed at resolving conflicts and thus enabling action. This is not a process driven by rational analysis or expert judgment, but by public debate over competing interests and values. Politics, as described by Harold Lasswell,

> is the process by which the irrational bases of society are brought out into the open....[It] is the transition between one unchallenged consensus and the next. It begins in conflict and ends in a solution. But the solution is not the "rationally best" solution, but the emotionally satisfactory one.[15]

Lasswell's use of the words "irrational" and "emotionally satisfactory" is not pejorative; rather, he is acknowledging the innate complexity and diversity of a dynamic society, and thus the irreducible messiness of the political process. People have legitimately different interests and perspectives that they will naturally attempt to protect and promote. Democratic politics gives them a forum for doing so. Not only is there nothing wrong with the consequent messiness, but all historical indications suggest that there is no viable alternative in a society that values freedom and justice and seeks to balance individual rights with the collective good. (Technocracy—rule by technical expertise—is not a viable alternative. As the history of the former Soviet Union demonstrated, technocracy is not only inherently authoritarian, it is even more irrational than democracy. This irrationality derives from the effort to impose rigid technical solutions on problems that reflect conflicting values and interests. The intrinsic diversity of human interests cannot be accommodated, so it must be suppressed.[16]

Thus, the goal of politics is the achievement, through democratic debate, of an operational consensus that enables action. This is a very different goal from that of science, which seeks to expand insight and knowledge about nature through an ongoing process of questioning, hypothesizing, validation, and refutation. Science progresses when it generates new questions more quickly than it resolves old ones, when it probes existing problems with increasing depth and acuity, when it uncovers new problems that were previously unrecognized, and when it reveals the limitations and failures of previous research. Good science is always pushing into the realm of the uncertain and the unknown. When a scientific problem is contentious and the object of a vibrant research effort, consensus is extremely difficult to achieve—the process of scientific investigation intrinsically militates against, and is designed to inhibit, premature consensus. Thus, if scientists are doing their job, then "more research" in the short term is invariably a prescription for raising new questions, problems, and uncertainties—for preventing, not achieving, consensus.

THE DEMANDS OF POLITICS VERSUS THE REALITY OF NATURE

Of course, in the search for a consensus that can enable action, what politicians would really like from science are facts to bolster irrefutably their partisan political positions. But the types of environmental problems facing policymakers are rarely if ever amenable to traditional reductionist approaches that yield unambiguous, statistically well-constrained answers. Rather, these problems are multivariate and nonlinear, and they comprise the behavior not only of evolving natural systems but also of humans. The desire to dispose of nuclear waste through geological isolation, for example, requires an understanding of the evolution and interaction of radiogenic, climatic, hydrogeologic, tectonic, volcanic, and social systems over a period of tens of thousands of years. The scientific certainty that politicians crave—that a repository will be "safe" over a given period of time—simply cannot be delivered. Uncertainties about the behavior of the repository are difficult to quantify, and they increase the further one looks into the future.

If one expands the scale of the problem from a relatively small area designated for nuclear waste disposal, to the entire global climate system, then these difficulties are compounded. The U. S. Global Climate Change Research Program promises a predictive capability that will enable decision making, but fundamental aspects of the climate system, ranging from the operation of the global carbon cycle to the behavior of the coupled atmospheric-oceanic system, are not yet adequately understood even in the present.[17] Moreover, when debate focuses on national, regional, or local impacts, the political stakes begin to rise rapidly, while at the same time the uncertainties associated with the predictive models begin to increase, because the details of a complex system behavior are much more difficult to characterize than are general attributes. To make matters worse, predicting future anthropogenic contributions to global climate change requires an accurate characterization not just of complex climate dynamics, but also of societal processes such as technological advance, economic development, and societal responses. As daunting as it is to understand and predict fully the physical and chemical evolution of the climate system, it is even more difficult to forecast societal trends.[18] Who, for example, foresaw in the mid-1990s that gasoline prices in the United States only a year or so later would be at a fifty-year low (adjusted for inflation), thus undermining any political willpower to control consumption to reduce greenhouse gas emissions?

Policy-making is an inherently forward-looking activity, and politicians naturally enlist scientists to provide predictions that can enhance foresight, and thus contribute to policy development. However, in the realm of complex environmental controversies, the capacity of science to provide predictive information that serves the needs of policymakers has yet to be demonstrated. Predictive models are invariably fraught with assumptions, simplifications, and judgment calls introduced by the modelers themselves, who must make a trade-off between real-world complexity and scientific tractability. From a scientific perspective, such trade-offs are both necessary and justifiable, but they open the models up to debate and criticism among experts, and skepticism from nonexperts. Whether assessing the number and size of old-growth forest plots necessary to preserve the endangered northern spotted owl over the next century, or estimating the extent of global temperature rise (not to mention its regional impacts on, for example, agricultural production), the scientific methods used to generate predictions have themselves all too often become a subject of political dispute, rather than an aid to resolving dispute.[19]

PRUDISH POLITICS, PROMISCUOUS NATURE

A final reason why science might not help resolve environmental controversies is less obvious, and perhaps more intractable, than the previous arguments. One may tend to think of human values as mutable and lacking clarity, especially in contrast to the fixed scientific laws of nature, and the supposedly rational and orderly process of scientific research. Yet the basic issues at stake in political debate—allocations of power and resources; trade-offs between justice and equality, between individual and community freedoms—do not in their essence change much over time. In contrast to the rather fixed array of human concerns underlying politics, nature's richness is sufficient to provide insights that can give comfort to all sides of the typical environmental policy debate.

Science itself is spectacularly diverse. Consider a scientist engaged in research in high-energy physics as part of a large research group, dependent on the technology of a huge and expensive particle accelerator, searching for indirect evidence of fleeting subatomic particles whose existence would support a mathematically derived theoretical explanation of the fundamental structure of the universe. What does this work have in common with the activities of a field geologist, mapping in a wilderness, alone or in a small team, recording direct observational data by hand in a notebook, trying to understand the paleoecology represented in a sequence of sedimentary rocks? Everything is different about these activities—the social, physical, and institutional setting of the work, the intellectual skills and methods that are used, the role of technology and direct observation, the standards of success and reliability, the manner in which accrued knowledge will be communicated, the potential utility of this knowledge, even—especially—the mental picture of nature itself.

Distinct scientific frames of reference often lead to distinct and not necessarily reconcilable bodies of knowledge about nature. For example, a geological perspective on the history and evolution of climate change yields an entirely different set of insights from those derived from atmospheric sciences. The geological view of climate is filtered and integrated by time: Geology sees the consequences of climate—recorded in the geological record—not the causes. Geologists seek to reconstruct ancient climate patterns and trends using proxies—indirect geological indicators of climate conditions—such as isotope variations in sediments and glacial ice; changing volume and composition in the dust content of polar ice; changes in abundance and morphology of planktonic marine microorganisms; and patterns of terrestrial paleobiogeography. Such proxies yield information on characteristics such as water temperature, rainfall abundance, atmospheric carbon, and sea level, but, more importantly, they demonstrate the intimate relation among atmospheric, biospheric, and lithospheric processes over history. The geological view is necessarily integrative and retrospective; it is capable of imaging climate variability at time scales unavailable to direct observation; it recognizes the extreme contingency, and thus indeterminacy, of the dynamic atmospheric system.

In contrast, the atmospheric science view focuses on characterizing the physical and chemical state and dynamics of the present-day atmosphere, and on modeling possible future changes. Atmospheric science pursues knowledge about the climate system by investigating causal relations among the innumerable components of the system. These components, which can often be directly measured (albeit sometimes with great difficulty), include everything from moisture gradient and aerosol content in the atmosphere to heat and water fluxes at the ocean–atmosphere interface to patterns and processes of carbon sequestration. The atmospheric science view has a fundamentally reductionist and deterministic component. It is rooted in the search for causation; it seeks to combine theoretical "first principles" that govern the climate system (mathematical representations of basic physical principles) with quantified observational data to yield predictive and "retrodictive" models[20] of system evolution.[21]

Geologists struggle to piece together a historical record of atmospheric change, but there is little that they can say about causation, because the details of the complex climate system have been erased by time. Atmospheric scientists, in contrast, are awash in detailed observation and bolstered by theory, but they can never validate their models because climate is an open system, and is therefore unpredictable.[22] The views achieved by these two approaches cannot necessarily be integrated. Atmospheric models, for example, have not been able to account for rapid destabilizations of the climate system that are seen in the geologic record, and they will never predict such changes with certainty.[23] The models can only reproduce the contingent conditions that have triggered such rapid change in the past if the scientists who design the models make ad hoc assumptions about the future—assumptions that would undermine the scientific credibility of the models.[24]

Which view is more correct: the record of contingency and long-term indeterminacy revealed through reconstruction of past climate, or the evidence of causal relations between measurable system components determined by observation, measurement, and theory in the present? The question is meaningless; these are equally valid perspectives on nature, yielding their own sets of scientifically objective insights.

The great diversity of scientific approaches to understanding nature suggests a similar or, more likely, much greater diversity in nature itself. In other words, nature can be studied and understood by science on many levels, because nature operates on many levels. To be blunt: Despite the insistence of many scientists and philosophers that all is reducible to physics,[25] there is no empirical basis for such an assertion—the weight of evidence is thus far firmly on the other side. Although the progress of science shows that disparate activities can indeed coalesce (as spectacularly seen in the case of molecular genetics, which arose out of organic chemistry and genetics), in many other cases, once-coherent disciplines undergo an irrevocable shattering. The life sciences, for example, are clearly sundered along a line that separates reductionist, molecular perspectives (e.g., molecular genetics) from macroscopic, systemic views (e.g., ecology). The proliferation of specialty journals in the sciences,[26] in part reflecting an academic culture that rewards specialization over synthesis, must certainly bespeak, as well, of the inherent Humpty-Dumptyness of nature.

Again, consider global climate change research. The USGCCRP encompasses nearly 100 different research projects at ten federal agencies in areas ranging from "global ocean ecosystem dynamics" to "impacts of climate change on energy fluxes."[27] What is the likelihood that such a wonderfully diverse program will yield a unified picture of global climate change that can generate and support progress toward a political consensus? More likely, it will lead to numerous and perhaps conflicting perspectives that can be invoked by policymakers to support various sides of the issue and conflicting policy prescriptions.

A vivid example of this problem emerged from a recent exchange among prominent scientists concerned about climate change. The debate was triggered when a "Scientists' Statement on Global Climate Disruption" was distributed by E-mail to numerous researchers, in an effort to collect signatures prior to distribution to the media. The essence of the "Scientists' Statement" is captured in the following excerpt:

> We are scientists who are familiar with the causes and effects of the climatic disruption summarized recently by the Intergovernmental Panel on Climate Change (IPCC). We endorse those reports and observe that the further accumulation of greenhouse gases commits the earth irreversibly to further warming and to further destabilization of global climate. The risks associated with such changes justify preventive action through reductions in emissions of greenhouse gases. As the largest emitter of greenhouse gases, the United States must take leadership by fulfilling its commitment to reductions in its emissions.
>
> Global climatic disruption is under way. The IPCC concluded that global mean surface air temperature has increased by between 0.54 and 1.08 degrees Fahrenheit in the last 100 years and anticipates a further continuing rise of 1.8 to 6.3 degrees Fahrenheit during the next century. Sea-level has risen on average 4–6 inches during the past 100 years and is expected to rise another 6 inches to 3 feet by 2100. Warmer temperatures cause an amplified hydrological cycle with increased precipitation and flooding in some regions and more severe aridity in other areas. The warming is expanding the geographical ranges of malaria and dengue fever and can be expected to open large new areas to human diseases and plant and animal pests. Effects of the disruption of climate are sufficiently complicated for us to assume that there will be effects not now anticipated.
>
> Our familiarity with the scale, severity, and costs to human welfare of the disruptions that the climatic changes threaten leads us to introduce this note of urgency and to call for early domestic action to reduce U.S. emissions....28

The names of six prominent scientists appeared on the statement as initial signatories. Shortly after the statement was distributed, Tom Wigley of the National Center for Atmospheric Research, a leading climate modeler, responded by E-mail:

> While I hold the [signatories] in high regard, I do not consider them authorities on the climate change issue.
>
> Phrases like (my emphasis) "climate DISRUPTION is under way" have no scientific basis, and the claimed need for "greenhouse gas emissions (reductions) begin-

ning immediately" is contrary to the careful assessment of this issue that is given in the IPCC reports.

No matter how well meaning they may be, inexpert views and opinions will not help. In this issue, given that a comprehensive EXPERT document exists, it is exceedingly unwise for highly regarded scientists to step outside their areas of expertise. This is not good scientific practice....29

John Holdren of the John F. Kennedy School of Government, an initial signatory, responded directly to Wigley's message:

Dr. Wigley's critique of the "6 scientists' statement" on global climatic disruption is surprising and, in all of its principal contentions, completely unconvincing....

Dr. Wigley has written that he does not consider the signers of the "6 scientists' statement" to be "authorities on the climate change issue" and that "Inexpert opinions do not help." Since he is a climatologist, one supposes that he would have been at least somewhat less distressed if a statement of this sort had been issued by members of that profession. Do they hold the only relevant "expertise"? What part of "the climate change issue" is he talking about here?

Understanding how the climate may change in the future, of course, depends on insights not only from climatologists but also from soil scientists, oceanographers, and biologists who study the carbon cycle; from energy analysts who study how much fossil fuel is likely to be burned in the future and with what technologies; from foresters and geographers who study the race between deforestation and reforestation; and so on....

Now, as it happens, the signers of the "6 scientists' statement"—whom Dr. Wigley deems not to be "authorities on the climate-change issue" and, indeed, so "inexpert" as to render an expression of their opinion "not helpful"—include an atmospheric chemist who shared the 1995 Nobel Prize in chemistry for his work on chlorofluoro-carbons and stratospheric ozone...; an ecologist widely recognized as one of the foremost analysts of the role of forests in the carbon cycle; two of the world's leading authorities on the structure, function, and vulnerability to disruption of the world's plant communities; a distinguished marine ecologist...; and (myself) an individual who has been studying for 30 years the local, regional, and global environmental impacts of the world energy system and the technical and policy options for meeting world energy needs in less damaging ways. Is our knowledge less relevant than Dr. Wigley's...to reaching a reasoned judgment on the seriousness of the climate-change issue and on what needs to be done about it? I think not.[30]

In response to this fusillade, Wigley expanded his critique in a second E-mail:

Let me point out some specific problems with the "6 scientists' statement."
1. They state that "global climatic disruption is under way." The word "disruption" occurs on a number of occasions in the text, and in the title. In the two dictionaries I have looked at, disruption is defined as "the act of rending or bursting asunder" and "throwing into confusion and disorder."

The above statement is incorrect. If you compare it with the IPCC statement that "the balance of evidence suggests a discernible human influence on global climate," you will notice a radical difference. The "6 scientists' statement" goes far beyond what IPCC says....This is not mere semantics: in issues like this, one must be very careful in one's choice of words....

2. They state that there is a need to produce "a substantial and progressive global reduction in greenhouse gas emissions beginning immediately"....I note that IPCC does not make any such statement and (more importantly) that the 6 scientists give no basis for their own categorical statement....

3. Item 1 [above] is in the area of climate data analysis. None of the 6 scientists has specific expertise in this area. Item 2 bridges the fields of carbon cycle modeling and economics. I do not think any of the 6 scientists have such expertise, although I admit that they, as a group, have some knowledge that impinges on carbon cycle modeling....[31]

Holdren's basic point—that climate change is the domain of many disciplines, all of which have important insights that can be brought to bear on the policy problem—is acute and indisputable, but does not overrule Wigley's argument that only climatologists can truly understand the intricacies of climate data analysis. They are talking about different things, Wigley adhering to a rather rigid definition of climate, Holdren referring in the broadest sense to the climate system and its interaction with biological and social systems. Each is an indisputable expert with an enthusiasm for brandishing his unimpeachable credentials in support of the legitimacy of his position. What, then, is a policymaker supposed to do?

7.3 AN EXCESS OF OBJECTIVITY

In other words, we are not suffering from a lack of objectivity, but from an excess of it. Science is sufficiently rich, diverse, and Balkanized to provide comfort and support for a range of subjective, political positions on complex issues such as climate change, nuclear waste disposal, acid rain, or endangered species.

This observation, if it is anything close to the mark, suggests that in the political arena, subjectivity and objectivity are not separate and immiscible realms that must always be kept apart, but rather that they are closely related attributes of any highly complex societal problem—opposite sides of the same coin. The science wars, by promoting a false distinction between "constructed" and "objective," have diverted attention from this fundamental problem. When an issue is both politically and scientifically contentious, then one's point of view can usually be supported with an array of legitimate facts that seem no less compelling than the facts assembled by those with a different perspective. In the midst of such controversy, the boundary between facts and values invariably becomes much fuzzier than we often make it out to be. The problem is not one of good science versus bad, or "sound" science versus "junk" science. The problem is that nature can be viewed through many analytical lenses, and the resulting perspectives do not add up to a single, uniform image, but a spectrum that can illuminate a range of subjective positions.

The above dialogue illustrates just this problem. To the climate modeler, a small, anthropogenic contribution to global temperature does not amount to climate "disruption," because the climate system is not fundamentally destabilized. To an ecologist, however, small temperature variations could stimulate significant changes in ecosystem function. The latter view might suggest the need for rapid policy action to control greenhouse gas emissions, even at high economic cost; the former might support a more cautious, less economically disruptive approach. Additional scientific insights add more complexity to the problem. Research on energy production shows that the United States is responsible for 24 percent of global carbon dioxide emissions.[32] This data could support the view that the nation has a responsibility to act decisively to limit emissions. Yet research on carbon cycling suggests that the United States may in fact sequester more carbon in its young forests than it emits from its massive industrial and transportation systems.[33] Such results could bolster an argument against U.S. action.

Or consider the example of acid rain. In the late 1970s and early 1980s, scientists, followed by environmental groups and the press, began reporting that acid rain was damaging many forests and lakes in the northeastern United States. Fingers were pointed at coal-burning power plants in the industrial Midwest, whose emissions were said to be the source of the problem. The politics of the controversy pitted the environmental and water quality of the Northeast against the economic health of the Midwest. Congress responded in the predictable fashion: A research program was created to determine the causes and assess the impacts of the problem, and to recommend actions—based on the science—to mitigate those impacts. Ten years and $600 million later, the National Acid Precipitation Assessment Program had generated copious quantities of excellent science on the causes and impacts of acid rain, but had failed to achieve any sort of consensus scientific view that could motivate a political solution to the problem. This failure was probably unavoidable—the issue encompasses so many different problems, from the costs of reducing power plant emissions, to the assessment of forest damage and its various causes (including natural soil acidity)—each with its attendant uncertainties—that there is simply no such thing as a "right way" to look at acid rain.[34] When a political solution was achieved, it reflected little of the knowledge gained from the research program, but instead made use of an economic tool—tradable permits for sulfur oxide emissions—that helped to allay the concerns of Midwestern lawmakers about adverse economic impacts of a more rigid emissions control scheme. Only when this political solution was implemented, as part of the Clean Air Act Amendments of 1990 (PL, 101–549) could a new role for science come into focus: to monitor the impacts and effectiveness of the policy decisions, and to provide feedback into a political process that had already decided upon a general course of action.[35]

The close linkage between the subjective and objective elements of environmental policy debate creates another, more insidious problem. If you were a policymaker, would you rather participate in a debate about the scientific aspects of a controversy, or about the interests and values that underlie the controversy? Arguing about science is a relatively risk-free business; in fact, one can simply mobilize the appropriate expert to do the talking, and hide behind the assertion of objectivity. But talking openly about values is much more dangerous, because it reveals what is truly at stake.

Again, global climate change exemplifies the point. Press coverage, congressional hearings and debate, proclamations by environmental groups and industry groups all focus on the science, and the science, as I have tried to show, can serve them all well. Hidden by this discourse are the underlying issues that drive the problem of climate change: the future economic path of the postindustrial world, population growth and distribution, patterns of land use, the distribution of wealth and resources among nations, and the vulnerability of poor nations to natural and anthropogenic hazards.

These very issues were conspicuously on display and just as conspicuously ignored in November 1998, when thousands of people converged in Buenos Aires, Argentina, to haggle over the details of an international agreement to control greenhouse gas emissions—an agreement that, even if widely adopted (which is unlikely) can make very little contribution to controlling climate change.[36]

Meanwhile, less than a week earlier and 6000 km to the northwest, Hurricane Mitch had killed more than 10,000 people in Central America while virtually wiping out the economies of Honduras and Nicaragua—impacts that could have been significantly reduced through effective emergency preparedness and land use planning. While anthropogenic climate change may or may not exacerbate the frequency and severity of future extreme weather events (a scientific question that will not soon be resolved), the indisputable fact is that such events are a historical and future reality, regardless of what climate change science reveals about their causes. But reducing the vulnerability of poor nations to natural disasters is not a politically attractive topic, and all sides of the global climate change controversy find it safer to fight over the science than the value-laden issues that the science conceals.

7.4 THE POLITICAL IMPLICATIONS OF A GEOLOGICAL VIEW OF NATURE

The prevailing mental model of how science can help resolve environmental controversies (Fig. 7.1) has intuitive appeal. Can we possibly imagine that scientific facts applied to political problems will not help to bring those problems to resolution? I have tried to show that we need to revise our intuitions, because we are demanding from both politics and science what they are least likely to deliver: rationally optimal decisions on the one hand; consensus over a diverse body of relevant facts on the other.

Our misplaced expectations for science derive in part, I believe, from an overly restrictive view of how science extracts truth from nature. This restrictive view assumes that the culmination of science is the ability to develop predictive hypotheses and theories through highly controlled experiments (real, or imagined). Experiment serves to hold nature's complexity in abeyance, so that nature can be parsed into its component parts and governing laws. This is the physics view, dominant in modern culture, and for very good reason: The character and quality of modern life are derived in no small part from the transforming impact of science and science-based technologies, which in turn reflects a perspective on nature and a method of research derived from the success of physics. But nature can be viewed from another angle that is no less scientific, which is to say, no less devoted to creating a true picture of what is really out there. This might be called the "geological view," and it recognizes that nature, as experienced by humans and as recorded in the lithosphere and cryosphere, is the evolving product of innumerable complex and contingent processes and phenomena, revealed through historical reconstruction, and through analogy with what we see in nature today.[37]

The physics view, when applied to policy-making, promises to relieve humans from responsibility by generating predictions that can dictate action. The geological view is more modest, offering insight into the importance of context and the limits of foreknowledge. The former makes freedom unnecessary; the latter renders it essential.[38]

If we look at nature from the geological perspective, then the appropriate role of science in politics may come into clearer focus. Diversity, change, and surprise are accepted as the normal state of affairs, and uncertainty is not viewed as a problem to be overcome, but instead as a reality to be embraced—a source of the richness in nature that is consistent with the human experience. From this perspective, science would not be viewed as an authoritative voice that can cure us of politics, but as a source of insight that can help us understand the inevitable constraints on our knowledge and foresight, and therefore point us toward policy approaches that favor adaptation and resilience over control and rigidity.

From such a perspective, what roles can science be expected to play in environmental policy? Of course, it can alert society to potential challenges and problems that lie ahead. In fact, the threat of stratospheric ozone depletion, acid rain, and global climate change were brought to public attention and political prominence in part through the work of scientists. But, once an environmental issue becomes politically contentious, the geological view of nature accepts that science itself can become an obstacle to action. At this point, the quest for a "rationally best" solution must be abandoned as absurd, and attention must focus on defining complex problems in ways that allow politics to arrive at solutions. Again, global climate change illustrates the point. A huge commitment of scientific, political, and diplomatic resources has been made to the negotiation of a comprehensive international treaty governing the reduction of greenhouse gas emissions. Although all parties to the debate argue that science must be the basis for action, a unified scientific view of the problem and its potential political solutions fails to emerge, and indeed becomes more elusive with time. An approach more in line with the reality of science and politics looks for areas of potential consensus centered around smaller, related or component issues—energy efficiency, pollution abatement, technology transfer, natural hazard mitigation, land use planning. Consensus—and beneficial action—in such areas is easier to achieve because it involves fewer entrenched interests, a greater degree of concreteness, a lower degree of political risk, a lower cost of action, and a reduced price to be paid in the event of inevitable mistakes.[39]

A second role for science in environmental controversies thus emerges: to help guide action *after* political consensus is attained. The standard, linear model of science and politics is thus turned on its head, as shown in Figure 7.2. Because consensus already exists, action can be taken along lines that all parties can more or less agree on—the problem of excess objectivity is at least partly allayed. Politics has been allowed to do its job, and science becomes a tool to help determine if implemented policies are working as intended and if progress is being made toward agreed upon political goals. Results from such research then can be used to refine and redesign policies and programs and assess future options. This monitoring and assessment function should form the central contribution of science to environmental policy, yet it has been severely neglected in the United States,[40] perhaps because it subordinates science to politics.

These two principal roles for science in environmental controversies—diagnosis and assessment—are consistent with a geological view of nature and eschew the unattainable goals promised by the physics view—foreknowledge and control. And in the United States there is evidence that the geological view is gradually taking hold. The

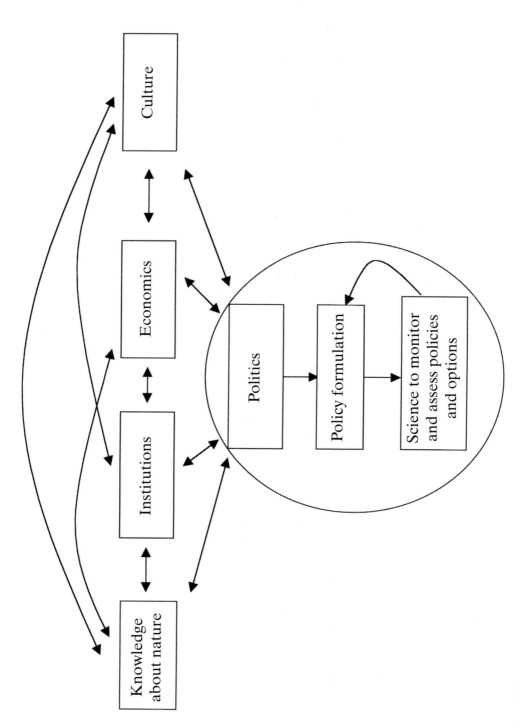

Figure 7.2 "Geologic view" of the relation among contingent variables in the search for solutions to environmental controversy. Scientific knowledge is one of many simultaneous inputs into the political process, but it cannot drive the process itself. The most important role for science comes after political consensus has been reached.

rise of participatory mechanisms for addressing environmental problems is beginning to replace top-down, command-and-control approaches that require authoritative knowledge as a precondition for action.[41] Participatory environmental decision making accepts that various stakeholders legitimately see reality in different ways; that there is no ultimate source of knowledge that can dictate the "correct" action under conditions of natural and societal complexity; and that the characterization of a problem, and consequences of any particular course of action aimed at addressing the problem, must always be uncertain. Policies are experiments; science can assess the success of the experiments and thus provide additional information that decision makers can integrate as they pursue longer-term goals. This feedback process is often called "adaptive management."[42]

In truth, adaptive management has become something of a buzzword in environmental policy circles, an idea with great theoretical appeal, much discussed and lauded but not yet proven in battle. This is not surprising. Frustration with the old approach to resolving environmental controversies demands change in the operation of political and scientific institutions—and such institutional change is always slow. A key step in this process requires the partial abandonment of a central tenet of modern society—the physics view of nature—and its replacement with a perspective that can encompass the multifaceted richness and diversity of reality as experienced by human beings—what I have here called the geological view. This view honors the reality of democratic politics and complex natural phenomena, and places science not outside this reality, but squarely within it.

NOTES

1. Brian Martin and Evelleen Richards, "Scientific Knowledge, Controversy, and Public Decision Making," in S. Jasanoff, G. Markle, J. Petersen, and T. Pinch (eds.), *Handbook of Science and Technology Studies* (London: Sage Publications, 1994), p. 513.

2. Bruno Latour and Steve Woolgar, *Laboratory Life: The Construction of Scientific Facts* (Princeton, NJ: Princeton University Press, 1986), p. 275.

3. Paul R. Gross, Norman Levitt, and Martin Lewis (eds.), *The Flight From Science and Reason* (New York: New York Academy of Sciences, 1996), p. 257.

4. Alan D. Sokal, "Transgressing the Boundaries: Towards a Transformative Hermeneutics of Quantum Gravity," *Social Text* 46/47 (Spring/Summer 1996): 960–968.

5. Stanley Fish, "Professor Sokal's Bad Joke," *The New York Times* (May 21, 1996): A13; Edward Rothstein, "When Wry Hits Your Pi From a Real Sneaky Guy," *The New York Sunday Times, The Week in Review* (May 26, 1996):6.

6. Paul Forman, "Assailing the Seasons," *Science* 276 (May 2, 1997): 750–752.

7. Gross, Levitt, and Lewis, *The Flight From Science and Reason.*

8. "You Can't Follow the Science Wars Without a Battle Map," *The Economist* (December 13, 1997): 77–79.

9. David L. Hull, *Science as a Process: An Evolutionary Account of the Social and Conceptual Development of Science* (Chicago: University of Chicago Press, 1988), p. 31.

10. Long before postmodernism was a gleam in the eye of French philosophers and sociologists, the German microbiologist Ludwig Flek wrote a beautiful and elegant book about the social construction of a single fact—the correlation between the Wasserman reaction and the presence of syphilis bacilli in the blood. See Ludwik Fleck, *Genesis and Development of a Scientific Fact* (Chicago: University of Chicago Press, 1979); originally published in German, 1935.

11. Committee on Earth and Environmental Sciences, *Our Changing Planet: The FY 1994 U.S. Global Change Research Program* (Washington, DC: National Science Foundation, 1993), p. 5.

12. Committee on Earth and Environmental Sciences, *Our Changing Planet: The FY 1998 U.S. Global Change Research Program* (Washington, DC: Office of Science and Technology Policy, 1998), p. 32

13. "Moratorium proposed on clean air rules," *Issues in Science and Technology* (Fall 1997): 33–34; Jocelyn Kaiser, "Showdown Over Clean Air Science," *Science* 277 (July 25, 1997): 466–469.

14. See Daniel Sarewitz, *Frontiers of Illusion: Science, Technology, and the Politics of Progress* (Philadelphia: Temple University Press, 1996), pp. 89–92.

15. Harold D. Lasswell, *Psychopathology and Politics* (Chicago: University of Chicago Press, 1977), pp. 184–185.

16. Loren R. Graham, *What Have We Learned About Science and Technology From the Russian Experience?* (Stanford, CA: Stanford University Press, 1998).

17. Curt Suplee, "Studies May Alter Insights Into Global Warming," *The Washington Post* (March 15, 1999): A7.

18. William Ascher, "The Forecasting Potential of Complex Models," *Policy Sciences* 13 (June 1981): 247–267; Jan Rotmans and Hadi Dowlatabadi, "Integrated Assessment Modeling," in Steve Rayner and Elizabeth L. Malone (eds.), *Human Choice and Climate Change: Volume 3. The Tools for Policy Analysis* (Columbus, OH: Battelle Press, 1998), pp. 291–377.

19. Steven L. Yaffee, "Lessons About Leadership From the History of the Spotted Owl Controversy," *Natural Resources Journal* 35 (Spring 1995): 382–412; Shiela Jasanoff and Brian Wynne, "Science and Decisionmaking," in Steve Rayner and Elizabeth L. Malone (eds.), *Human Choice and Climate Change: Volume 1. The Societal Framework* (Columbus, OH: Battelle Press, 1998), pp. 59–74.

20. That is, models designed to reproduce past atmospheric behavior in order to validate underlying assumptions.

21. Kevin E. Trenberth, *Climate System Modeling* (Cambridge: Cambridge University Press, 1993).

22. Jasanoff and Wynne, "Science and Decisionmaking."

23. W. S. Broecker, "What If the Conveyor Were to Shut Down? Reflections on a Possible Outcome of the Great Global Experiment," *GSA Today* 9 (January 1999): 1–7.

24. N. Oreskes, K. Shrader-Frechette, and K. Belitz, "Verification, validation, and confirmation of numerical models in the earth sciences," *Science* 263 (February 4, 1994): 641–646.

25. For a recent example, see Edward O. Wilson, *Consilience: The Unity of Knowledge* (New York: Alfred A. Knopf, 1988).

26. David L. Goodstein, "After the Big Crunch," *The Wilson Quarterly* 19 (1995): 53–60; Derek J. de Solla Price, *Little Science, Big Science* (New York: Columbia University Press, 1963).

27. Committee on Earth and Environmental Sciences, *Our Changing Planet: The FY 1998 U.S. Global Change Research Program*, pp. 77–107.

28. Jane Lubchenco, F. Sherwood Rowland, Peter H. Raven, John P. Holdren, George M. Woodwell, and Harold A. Mooney, initial signatories, "Scientists' Statement on Global Climatic Disruption," June 1997 (circulated for comments April and May 1997); available online at www.ozone.org.

29. April 30, 1997 message.

30. May 2, 1997 message.

31. May 5, 1997 message.

32. *World Resources 1998–99* (New York: Oxford University Press, 1998), pp. 344–345.

33. S. Fan, M. Gloor, J. Mahlman, S. Pacala, J. Sarmiento, T. Takahashi, P. Tans, "A Large Terrestrial Carbon Sink in North America Implied by Atmospheric and Oceanic Carbon Dioxide Models," *Science* 282 (October 16, 1998): 442–446.

34. Charles Herrick and Dale Jamieson, "The Social Construction of Acid Rain," *Global Environmental Change* 5 (1995): 105–112.

35. Richard A. Kerr, "Acid Rain Control: Success on the Cheap," *Science* 282 (November 6, 1998): 1024–1027.

36. Robert M. White, "Kyoto and Beyond," *Issues in Science and Technology* (Spring 1998): 59–65; Rob Coppok, "Implementing the Kyoto Protocol," *Issues in Science and Technology* (Spring 1998): 66–74.

37. Robert Frodeman, "Geological Reasoning: Geology as an Interpretive and Historical Science," *Geological Society of America Bulletin* 107 (August 1995): 960–968; Victor R. Baker, "The Geological Approach to Understanding the Environment," *GSA Today* 6 (March 1996): 41.

38. An analogous and equally valid argument could be made for the differing perspectives of ecology and molecular biology.

39. Ronald D. Brunner, "Global Climate Change: Defining the Policy Problem," *Policy Sciences* 24 (1991): 291–311.

40. Committee on Environmental Research, *Research to Protect, Restore, and Manage the Environment* (Washington, DC: National Academy Press, 1993).

41. Marc K. Landy, Megan M. Susman, and Debra S. Knopman, *Civic Environmentalism in Action: A Field Guide to Regional and Local Initiatives* (Washington, DC: Progressive Policy Institute, 1999); Kai Lee, *Compass and Gyroscope* (Washington, DC: Island Press, 1993); Lisa Jones, "Howdi Neighbor!

As a Last Resort, Westerners Start Talking to Each Other," *High Country News* (May 13, 1996): 1, 6, 8.

42. Best defined in Lee, *Compass and Gyroscope*, and L. H. Gunderson, C. S. Holling, and S. S. Light (eds.), *Barriers and Bridges to the Renewal of Ecosystems and Institutions* (New York: Columbia University Press, 1995).

8

The Transparency and Contingency of Earth

Albert Borgmann

Albert Borgmann has been professor of philosophy at the University of Montana for more than 30 years. Stimulated by his persistent attempt to live and think in a land of high mountains and big sky, there has emerged what may be called a small Borgmann school in contemporary thought. Borgmann's work includes Technology and the Character of Contemporary Life: A Philosophical Inquiry *(Chicago: University of Chicago Press, 1984), and most recently* Holding On to Reality: The Nature of Information at the Turn of the Millennium *(Chicago: University of Chicago Press, 1999).*

The idea with which Borgmann is most often associated is the distinction between what he calls "things and devices." An archetypal thing, for Borgmann, is a wood heating stove, the function of which is clearly transparent and which easily becomes a center for human life. By contrast, the thermostatically controlled central heating system is a device that hides its inner workings and never stands out as significant in human interactions—unless, of course, it breaks down. In the world of things, humans orient themselves by realities other than themselves; in the world of devices, it is human wants and interests that come to the forefront as orienting principles, producing the consumer society.

In the essay below, Borgmann distinguishes between two tasks of geology, what he calls the "scientific" and the "disclosive." The scientific task seeks to account for geologic facts by subsuming these facts under the laws of physics and chemistry. In contrast, the disclosive task of geology seeks to reveal the ways of natural processes so that we can adjust ourselves to them, and by doing so come to be at home in the world. For Borgmann, such disclosure is fundamentally local in nature, revealed to us through understanding the characteristics of the landscape that we inhabit. Disclosure geology helps us regain a sense of reverence and depth missing in our lives.

Geology is entrusted with the study of the very ground whereon we walk; and being so fundamental a science, it is, not surprisingly, charged with basic tasks. It warns us of earthquakes. It informs us about the availability of groundwater, of fossil fuels, and of ores of various kinds. It tells us whether and how a construction site will support a building or a structure of civil engineering. These and similar responsibilities are as evident as they are crucial to our welfare. I will call them the "proximate tasks" of geology to distinguish them from the "profounder tasks" that are even more important to human well-being, if much less evident.

The profounder tasks come first into view when we consider that human beings pursue their daily tasks within a background of experiences and assumptions. This background has a spatial and temporal dimension. As I write this, I have it in the back of my mind that this is a room in a building at the University in Missoula, a town in Montana, a state of the United States, a country on the North American continent, and so on. And, I am constantly if implicitly aware that this is the autumn in the second to the last year (as we count them) of the second millennium.

In one regard my spatial and temporal background knowledge has a definite bearing on what I am centrally concerned with. Should I learn that I am badly mistaken about my background, I would be thoroughly disoriented, and the endeavors in the central area of my attention would fall apart. Thus, if someone were to convince me that I am actually in Laramie, Wyoming, or that I had only another week left to live, I would be at loose ends and stop what I am doing.

In another respect, however, it is quite unclear what conception people have of their ultimate horizons, how well their conception accords with what science tells them about these encompassing backdrops, and what bearing their knowledge has on their sense of the world and on their conduct. Presumably we know so little about those things because there is not much to know. Conversations with my students, at any rate, suggest that the common cosmic awareness is vague and often inconsistent where it is detailed.

This kind of unconcern and ignorance is not new, but it has lately been acquiring a peculiar fluorescence due to developments in information technology. We have loosed a flood of easily available and newly glamorous information. Our ability to surf this ocean of information effortlessly has given us a sense of lightness that seems to defy the gravity and density of reality as we once knew it. Lightness has its dark side, however. Lack of density leads to perilous bloating and lack of gravity to aimless floating. We may be losing our ability to be at home on Earth. To counter the vacuity and mindlessness that threaten our culture we may do well to turn to a discipline that appreciates density and gravity.

In fact, geology is in part the successor of those disciplined human endeavors that sought to help us understand the planet we live on. Early endeavors took the form of myth. The one that at least etymologically stands at the beginning of geology tells us about Gaia, Mother Earth. The story is part of Greek mythology and has been recounted by Hesiod.[1]

> In the beginning there was only Chaos, the Abyss
> But then Gaia, the Earth, came into being
> Her broad bosom the ever-firm foundation of all,
> And Tartaros, dim in the underground depths,
> And Eros, loveliest of all the Immortals....
> Earth's first child was Ouranos, starry Heaven,
> Just her size, a perfect fit on all sides.
> And a firm foundation for the blessed gods.
> And she bore the Mountains in long ranges, haunted
> By the Nymphs who live in the deep mountain dells.
> Then she gave birth to the barren, raging Sea....

This myth, and many others like it, explains Earth by means of a story. In this it does not differ from historical geology. It too explains the world by telling the Earth's story. The crucial difference between the geological and the mythological tale is that the former tests and illuminates its accounts by placing them in the framework of scientific lawfulness. Geology as a science developed along with the natural sciences and gained in explanatory power as the lawfulness of science became more powerful and sophisticated. In physical geology, events and objects are subsumed under the laws of physics and chemistry. That is what makes geological explanations scientific. If a conjecture fails to fit into the framework of laws it is discarded or shelved. Thus, the claim that the water of mountain springs is forced up from the sea and cleansed of salt does not accord with hydraulics and chemistry and was refuted and rejected.

The lawfulness that frames Hesiod's account is weak and limited—it is the law of mythic birthing. Hence, Hesiod's explanation cannot be called scientific. But it has explanatory power of a different sort. It discloses the fertility of Earth, her sheltering solidity, her distinction from the dark underground, her kinship with Eros, and her mothering of heaven, the mountains, and of the sea. A disclosive explanation tells you what the crucial dimensions and forces of the world are, and it tells you or at least suggests how to find your place and come to be home in the world. Thus, the profounder task of geology is that of explaining Earth, and the explanatory task divides into two: the scientific and the disclosive.

Geology took the path of scientific rather than disclosive explanation. Propelled by sheer curiosity and guided by scientific lawfulness, it took on and answered the outstanding question about Earth: How many kinds of minerals are there? What are they composed of? What is the origin of mountains, valleys, and plains? What is inside Earth? How old is Earth? Where does the water of springs come from? What causes volcanoes to erupt? How did the continents get their shape and location?

Answers to these and similar questions have been accumulating over roughly four centuries, beginning in the sixteenth century with the work of Georgius Agricola, and coming to a grand synthesis and global conclusion in the late 1960s and early 1970s through the theory of plate tectonics. Within this framework, however, there is an infinity of details awaiting explanation. It seems like sheer good luck that, along with this problematic plethora, the means to cope with it have developed as well. Information technology and its allied disciplines have provided for entirely new and powerful instruments of gathering data and for the storage and processing capacities required by the immense amounts of data being collected. Geology by the early 1970s had rendered the history and structure of Earth surveyable in its major outlines. The goal now is to make Earth scientifically transparent. To make it fully so is an unreachable goal. Yet transparency is a powerful regulative idea that can be realized at least in parts and increasingly so. An example of how rock and sand can become transparent is the computer model of a piece of the Rabbits Hill oilfields in eastern Montana.[2] The model allows you to navigate through the layers beneath the oilfield and to look at its structure from various points and angles.

Dynamic models are yet more illuminating because they allow one to render transparent not only what lies beneath the surface but also what lies in the future. Yet the tendency toward transparency has revealed recalcitrance as well. Reality defies models as often as or perhaps more often than it obliges them. Precisely under the assault of the most powerful and sophisticated instruments and procedures ever, reality has shown itself to be endlessly and elusively complex. The problem is analogous to fitting a curve to data points. Given the complexity of minerals and water in the ground, the available data tend to underdetermine the rocky and watery reality to begin with. Sometimes the original model (the curve) needs to be tweaked repeatedly to fit present data. If a reasonable fit is achieved, there is never a guarantee that the correct model or curve has been found. Two (or more) models (or curves) may reasonably fit all data of the past and present and yet take very different trajectories for the future.[3] Ironically, when a superabundance of data becomes available from remote sensing devices, the data refuse to disclose of themselves the significant features of reality. Painstaking expertise is needed to firm up boundaries and highlight differences in the fuzzy display of data.[4] While in the modern era of science one had to be impressed by the lawfulness of reality, the dawning postmodern era impresses on us the unsurpassable contingency of reality.

Contingency can be taken negatively as the factual residue that stubbornly refuses to submit to theories or models. Stubborn resistance, however, may be eloquence in disguise. In contingency we can see the face of reality that we need to acknowledge rather than master. Faces, unfortunately, speak with less definition than do theories and factual findings. The meaning of a face can easily be mistaken or entirely overlooked. What is needed then is the disclosure of its meaning. In short, to be equal to the contingency of reality we need disclosive rather than scientific explanations. This comes to recovering that part of cosmic explanation that mythology had in abundance and was increasingly (and properly) given up by geology in its pursuit of scientific explanations.

Turning to disclosure is to take up the other of the profounder tasks of geology. It is more difficult and less conclusive, but at the same time it answers a deeper need— the need for a grounded understanding of the world in which we live. Geology, concerned with scientific explanations, needs the complement of a normative geology, dedicated to disclosive explanations.

Normative considerations raise red flags. They threaten to involve us in interminable and inconclusive squabbles. Hence, it seems advisable to stay in the objective realm of scientific geology. It needs to be stressed that there are weighty matters of fact in geology that cannot be dislodged by exposing hidden value judgments. The continents or rather the plates they are a part of have drifted whether we find that morally or socially agreeable or not. Moreover, some value assumptions are so well-taken and widely shared that they are not in dispute. Changing the approach to Earth from piety to curiosity was of course an epochal and fundamental normative move for geology. But once it had been raised by the pre-Socratic philosophers and after it had been installed as the ruling attitude by the Enlightenment, curiosity was not to be denied. It would have been cowardly and bigoted to evade it.

Geologists are proverbially curious. They are inveterate lifters of veils and concealments, and scientific curiosity was a necessary and sufficient guide to the large and sundry features that converged in the global closure of plate tectonics. No doubt there are many open questions where pure curiosity will continue to justify the pursuit of answers. Yet given the infinity of possible approaches and the growing flood of data, some sort of cognitive interest is needed to provide direction for research. A cognitive interest is nothing but a normative commitment, and wherever you find cognitive interests in geology, it is too late to close the door to norms. They are already inside geology.

Economic and environmental geology are two areas where the guiding norms are evident. The norms, moreover, have not emerged from within geology but have been imported into geology. As regards economic geology, I doubt that the percentage of geologists employed in the petroleum industry reflects the intrinsic interest petroleum commands in geology. Similarly for environmental geology. Compared to the upheavals and catastrophes Earth has seen over 4.5 billion years, the disturbances of the last hundred years are unremarkable. In both cases, however, a human norm, legitimate within limits, has been brought to geology, a norm of prosperity in economic geology and a norm of biological welfare in environmental geology.

Norms are necessary to give research direction. They are needed even more to meet the profounder disclosive task of geology—to give humans a sense of the world they inhabit. When modern geology began to make Earth in its depth and extent surveyable, it clearly aided human beings in their comprehension of the basic structure of their world. But as the realm of geological explanation became more expansive and transparent, it ceased to be inhabitable by human imagination. The geological account of terrestrial history is too vast to be appropriated by the human mind and body. As John McPhee has it, "[n]umbers do not seem to work well with regard to deep time. Any number above a couple of thousand years—fifty thousand, fifty million—will with nearly equal effect awe the imagination to the point of paralysis."

If paralysis is the bane of temporal expanses, fatigue is the curse of geological variety. McPhee tried to control the countless geological events on the American continent by selecting those that lay along today's Interstate 80. But especially in the mountainous regions of this chain, the heaving, crushing, subsiding, metamorphosing, eroding and more occurs so many times that soon we feel unequal and exhausted when once more our astonishment is called upon to do justice to a cataclysm.

When it comes to grounding our comprehension of the world, all geology must be local. Normally, the place that we inhabit already has a face we comprehend and hold dear. Moreover, the physiognomy of the landscape is what geology has come to. Earth's geological structure has literally and figuratively fractal features. Coastlines and bare mountains can be literally fractal. From afar they display patterns of delineation or thrusting. Upon closer inspection they show the same patterns at a larger scale, and so on ad infinitum, at least in the mathematically idealized case. A landscape is figuratively fractal. It is shaped by large events, like continental drift. Layered on these are smaller events such as lifting and folding, and on those rest yet smaller events like the last glaciers that scoured the valley and left their moraine. But unlike a truly fractal structure, a landscape is given—as far as we, its inhabitants, are concerned—a final shape and character.

To the untutored and careless, the givenness of a landscape is brute and opaque, and their inhabitation of the place is thoughtless accommodation to its grade, composition, and soils. Disclosive geology, however, lifts the veil of unconcern and allows us to see how the land once moved and has now come to rest. A penetrating view of the place we inhabit allows us to deepen our life by adjusting the pace and rhythm of our enterprises to the shape and movement of the land.

In Norman Maclean's "A River Runs Through It," the life that is told and the story that tells it register and reply to the geology of the Blackfoot River.[6] At the start of the story, the narrator and his brother see their extraordinary vigor and prowess reflected in the straight and powerful course of the river. "Paul and I," says the narrator

> fished a good many big rivers, but when one of us referred to "the big river" the other knew it was the Big Blackfoot. It isn't the biggest river we fished, but it is the most powerful, and per pound, so are its fish. It runs straight and hard—on a map or from an airplane it is almost a straight line running due west from its headwaters at Rogers Pass on the Continental Divide to Bonner, Montana, where it empties into the South Fork of the Clark Fork of the Columbia. It runs hard all the way.[7]

Yet the river owes the particulars of its straightness and drop to a catastrophic event, "the biggest flood in the world for which there is geological evidence; it was so vast a geological event that the mind of man could only conceive of it but could not prove it until photographs could be taken from earth satellites."[8] The passage reminds us that through geology we come face-to-face with powers that dwarf whatever strength and skill we may claim for ourselves. Maclean's remarks also teach us humility in demonstrating that to appreciate geology we need to respect its details. That there once was a Lake Missoula was established by Joseph Thomas Pardee in 1910.[9] That there has been an immense flood racing across the Columbia Plateau in eastern Washington was discovered by J. Harlen Bretz in the 1920s.[10] In 1942, long before satellites were launched, Pardee determined that Lake Missoula was the source of the flood.[11] There were, in fact, multiple floods, as Victor R. Baker definitively demonstrated in 1973,[12] too late perhaps to be noted in Maclean's story, which was published only three years later.

Reverence is not a norm that needs to be injected into geology. It arises of itself from geology and makes it inhabitable. Though reverence entails devotion to truth, it does not imply mastery of the field. Local knowledge should be undergirded by a global knowledge of geology, but it can be selective and respond to geology where the contingencies of life suggest it. Maclean's story approaches its bittersweet climax when the narrator, his father, and brother go fishing for one last glorious time. This is a moment of high contingencies—crucial strands of life converge and contact one another. So do disparate forces of geology. The narrator stops the car on the way up the Blackfoot, and they look down to the river.

> We stopped and peered down the bank. I asked my father, "Do you remember when we picked a lot of red and green rocks down there to build our fireplace? Some were red mudstones with ripples on them."
>
> "Some had raindrops on them," he said. His imagination was always stirred by the thought that he was standing in ancient rain spattering on mud before it became rocks.
>
> "Nearly a billion years ago," I said, knowing what he was thinking.
>
> He paused. He had given up the belief that God had created all there was, including the Blackfoot River, on a six-day work schedule, but he didn't believe that the job so taxed God's powers that it took Him forever to complete.
>
> "Nearly *half* a billion years ago," he said as his contribution to reconciling science and religion.[13]

The recent ravages of the floods contrast with the ancient reassurance of the Precambrian rocks that the Macleans came to inhabit.

Reverence for geology breaks through the vacuous shallowness of contemporary life and restores a sense of depth. Once attuned to the profounder dimensions of reality, the world of rocks becomes not only inhabitable but also audible. It is the instruction of the river that first inspires the narrator to remember and tell the story of his family.

> It was here, while waiting for my brother, that I started this story, although, of course, at the time I did not know that stories of life are often more like rivers than books. But I knew a story had begun, perhaps long ago near the sound of water. And I sensed that ahead I would meet something that would never erode so there would be a sharp turn, deep circles, a deposit, and quietness.[14]

As you may know, the story of the Macleans ends unhappily. His parents and his brother are unable to save Paul from his rush into self-destruction. Within a year, drinking, gambling, and fighting led to his death. The narrator has to tell his parents that "my brother had been beaten to death by the butt of a revolver and his body dumped in an alley."[15] Yet the sense of reverence and depth that sustains father and brother allows them in the end to accept Paul's fate. When the narrator in his old age gives us an account of his family, there is in fact more than acceptance. In the concluding passage there is homecoming and consolation.

> Eventually, all things merge into one, and a river runs through it. The river was cut by the world's great flood and runs over rocks from the basement of time. On some of the rocks are timeless raindrops. Under the rocks are the words, and some of the words are theirs.
> I am haunted by waters.[16]

Geologists are dedicated to public education. So there is no need for reform or a new beginning. At most there may be room for orientation, confirmation, and a few suggestions. Let me try to summarize these. At this juncture of contemporary culture, society at large is imperiled by a sort of lightheadedness. Geology is not primarily or uniquely required to counter this malaise. It is a task for many hands and heads. But geology can certainly make a distinctive contribution to this enterprise. In reclaiming its ancient disclosive and normative heritage, geology can help us regain a sense of reverence and depth. Such a sense, in turn, will make geology inhabitable for laypeople, provided education in geology proceeds, within a roughly sketched global framework, in response to local contingencies. This can be done in the soberly helpful way that my colleagues David Alt and Don Hyndman have championed or in the profoundly eloquent way of Norman Maclean's "A River Runs Through It."[17]

Finally I want to suggest, with far less confidence and competence, that a sense of reverence for the depth of contingency may be helpful to geologists in avoiding servility to commercial or ideological norms and futility in the aimless pursuit of transparency where inquiries are undertaken chiefly because they can be undertaken. Here too a helpful paradigm is at hand. Geology, or at least paleontology, can be more than joyless cataloging or mindless Darwinizing. It can celebrate the fact that, as Stephen J. Gould has stated it, there is something like wonderful life.[18]

NOTES

1. Hesiod, *Works and Days and Theogony*, trans. Stanley Lombardo, ed. Robert Lamberton (Indianapolis: Hackett, 1993), p. 64.

2. *DOE Petroleum Reservoir Project* (Online: Montana Organization for Research in Energy). Available: http://ftp.cs.umt.edu:80/DOE/home.html. 10 June 1998.

3. Naomi Oreskes, Kristin Shrader-Frechette, and Kenneth Belitz, "Verification, Validation, and Confirmation of Numerical Models in the Earth Sciences," *Science* 263 (4 February 1994): 641–46.

4. J.-C. Muller, "Generalization of Spatial Databases," *Geographical Information Systems*, ed. David J. Maguire, Michael F. Goodchild, and David W. Rhind (Harlow: Longman, 1991), pp. 457–75.

5. John McPhee, *Basin and Range* (New York: Farrar, 1981), p. 20.

6. Norman Maclean, *A River Runs Through It and Other Stories* (Chicago: University of Chicago Press, 1976).

7. *Ibid.*, p. 12.

8. *Ibid.*, pp. 12–13. See also p. 83.

9. Joseph Thomas Pardee, "The Glacial Lake Missoula," *Journal of Geology* 18 (1910): 376–86.

10. Victor R. Baker, ed., *Catastrophic Flooding: The Origin of the Channeled Scabland* (Stroudsburg, PA: Dowden, 1981), pp. 20–58.

11. *Ibid.*, pp. 169–81.

12. Victor R. Baker, "Paleohydrology and Sedimentology of Lake Missoula Flooding in Eastern Washington," *Geological Society of America Special Papers*, no. 144 (1973).

13. Maclean, *A River Runs Through It*, p. 84.

14. *Ibid.*, p. 63.

15. *Ibid.*, p. 102.

16. *Ibid.*, p. 104.

17. E.g., David Alt and Don Hyndman, *Roadside Geology of Montana* (Missoula, MT: Mountain Press, 1986).

18. Stephen Jay Gould, *Wonderful Life* (New York: W. W. Norton, 1989).

9

Natural Aliens Reconsidered:
Causes, Consequences, and Cures

Max Oelschlaeger

Max Oelschlaeger is the Frances B. McAllister Endowed Chair in Community, Culture, and Environment at Northern Arizona University (NAU). His work at NAU focuses upon his creation of a Public Humanities Program that seeks to restore a place for the humanities within community life. His recent books include the Pulitzer Prize-nominated The Idea of Wilderness *(Yale University Press, 1991) and* Postmodern Environmental Ethics *(State University of New York Press, 1995).*

In this essay Oelschlaeger argues that the cultural models that dominate today—our ideas concerning political economy, science and technology, philosophy and religion—deny that humankind is of and about the Earth. This denial has inner (e.g., psychological) and outer (e.g., ecological) consequences. Examples of this include the land perceived as and governed by policies that view it as a commodity, and our psyches being held in the clutches of materialism and consumerism. Despite our culture's embrace of these assumptions, they are antagonistic to human flourishing and indeed to the flourishing of other life on Earth.

Oelschlaeger is guardedly optimistic about the possibility of overcoming these assumptions. He finds provisional signs of change, as our attitudes concerning the economy, politics, and the nature of the human psyche shift toward reintegrating human beings with nature. Our chances of overcoming this alienation from nature will be improved through a dialogue between Earth scientists and ecophilosophers. Both groups have a deep appreciation of the importance of temporality, and together they offer the promise of resolving the temporal incongruities that plague the relationship between civilization and nature.

My goal here is to explore the ground where the Earth sciences and philosophy, especially ecological philosophy (i.e., a philosophy that takes "Earth matters" seriously), might come together.[1] I believe that there is a convergence between the Earth sciences and philosophy—one that is profoundly relevant to the pressing environmental issues confronting culture. This essay develops that thesis. One immediate way to advance my claim is by calling attention to the two senses of "Earth matters" suggested by enclosing the phrase in scare quotes. On the one hand, "Earth matters" refers generally to the biophysical contexts/processes (the evolved biophysical planetary system, such as the continents and mountains, the oceans and atmosphere, the carbon cycle and hydrological cycle), which frame life on Earth. More specifically, these are the Earth matters that the Earth sciences, broadly conceived, have traditionally studied. On the other hand, "Earth matters" points toward the sense that Earth is consequential, that the biophysical contexts and/or processes that sustain life count, if for no other reason than the survival of our own species. Earth also matters if eco-catastrophes are to be avoided, such as an anthropogenic (human engendered) mass extinction of life, an event that would take some 15 million years, minimally, from which to recover.

We live, however, in a cultural context, increasingly a globalist context, predicated on the denial of "Earth matters." In the first place, Earth is largely conceptualized as simply raw material, as a supply of resources for human economic appropriation, as an unlimited source of matter-energy that can, through industry and technology, be converted into an unending, cornucopian stream of consumer goods. Earth sciences, traditionally, are conceptualized as serving that agenda, by finding subterranean water resources, vast fields of crude oil, rich veins of precious metals, and the like. Second, the notion that Earth matters is seemingly trumped by an economics that serves narrowly human interests, a politics that serves primarily propertied interests and individual rights, and a philosophy of science that views causal control and technological domination of Earth as the primary aim. Economics specifically, and political-economy more generally, over-determines the processes of cultural production and reproduction; thus, Earth sciences and ecophilosophy have largely been confined to the cultural sidelines.[2] However, there is some chance that a dialogue between Earth scientists and ecophilosophers might collectively give voice to a fully robust notion that Earth matters.

One reason for trying to have such a conversation is that academic philosophy has too long been isolated from not only the serious consideration of the planet, which our species increasingly, if only temporarily, dominates, but from culture itself. Another reason is that Earth scientists and ecophilosophers have different perspectives that might, though a collaborative discourse, prove useful in helping society find its way toward sustainability—however vague that term might be.[3] As the new millennium turns, we can clearly see that the past century has been dominated by economic thinking. Just as clearly, as the reality of global ecocrisis becomes more and more evident, we can see that a dialogue concerning Earth matters has the potential to lead culture in directions where economic success does not also entail planetary ruin.

There is an initial problem, however, that must be faced before the conversation can be enjoined. We live in an age of academic specialization so intense that even members of the same discipline can often not talk across their subdiscipline. How am I, by training a philosopher, to speak to geologists and other Earth scientists? And how are they to understand me? My problem is compounded by the fact that most of my geology is the roadside, backpacking, day-hiking variety. My ecology is little better. And our problem is compounded by the fact that most scientists are so busy doing geology, ecology, and so on that few have time to read any philosophy, let alone environmental philosophy. My difficulty is further compounded by an editorial requirement that brings terror to the heart of the academic writer: to write an essay so that the general public and undergraduate students might comprehend my meaning, yet to do so in a way that does not sacrifice intellectual substance for a "readerly" style.

The upshot is that Earth scientists and philosophers as well will likely find themselves on strange ground. Earth scientists will find here neither discussion of notions like plate tectonics and continental drift nor accounts of technologies like remote sensing and geographic information systems. Philosophers will find neither mention of terms like the phenomenological *epoche*, epistemology, and hermeneutics, nor discussion of esoteric treatises like Ludwig Wittgenstein's *Tractatus* or Martin Heidegger's *Being and Time*. The good news is that some of the brightest people I know, and also the people most prepared to act on the basis of scientific insight and philosophic reflection, are those undergraduates and members of the public who believe that humankind teeters on the brink of multiple eco-catastrophes.

9.1

Because philosophy and geology have involved themselves throughout their histories with the concept of time, albeit in different ways and at different scales, I shall begin with some remarks on time itself, especially on the *temporal incongruities* that lie at the root of environmental dysfunctions. At the highest level of generality, we might describe the problem as a disjointedness between the temporal horizons that dominate civilization and those that frame biophysical processes. We culturally define time along economic, political, and psychological continuums, whereas natural horizons of time are conceptualized as geological, biological, and ecological continuums. Clearly, natural and cultural scales of time are different.[4] Politicians, for example, are elected on the basis of four, plus or minus two years, and they make decisions of environmental policy based primarily on electoral considerations. But floral and faunal species, alpine and boreal ecosystems, and atmospheric and hydrological cycles exist naturally on vastly different temporal scales.

To take another example, economists dominate our cultural conversations about time, relentlessly emphasizing quarterly profit-and-loss statements, monthly rates of employment, inflation, or economic growth, and daily fluctuations in the stock market. Yet "good economics," such as maximizing economic growth (that is, economic throughput, the conversion of natural resources through industrial production into desired consumer goods), although good for the stock market (investors) and pocketbook (consumers) in the short term, increasingly appears to be "bad biophysics," using nonrenewable resources produced over millions of years at unsustainable rates, and generating pollutants that perturb systems, such as the ocean and atmosphere, that evolved over millions of years.[5] Because the Earth sciences operate at different temporal scales than do politicians and economists, it could be that they have much to teach the human species about time, especially about the deep, deep past and the evolutionary processes undergirding all life.

I must note that there is a fine irony here, for our conceptions of geological, biological, and ecological time themselves exist within a civilizational framework, constituted primarily by professional communities that are themselves subject to the influence of psychological, economic, and political factors. The very notion of geological time, that is, time on the order of millions or tens of millions of years, did not exist as recently as two centuries ago. The nineteenth century is perhaps as much marked by the emergence of the notion of geological as biological time. But in any case, that is another issue for another time.

My point here is simply, as Shakespeare suggests, that *time is out of joint*. Within the cocoon of civilization we are dominated by senses of time emphasizing the short term. Geological time is of a different, unfamiliar, indeed, strange order. To put the idea in the words of a geologist: "The concept of geologic time is practically beyond human comprehension. We tend to think of time in terms of only a few days, weeks, years, or even decades, for these constitute a lifetime. The events of a few hundred years ago are considered ancient history, and it is almost impossible for us to conceive of a time a million, a hundred million, or…a billion years before now."[6]

Another way of putting the point is in terms of the solar system, some 5 billion years old, and life, some 3.8 billion years old, as over against the longevity of the human species, fewer than 500,000 years, and Western civilization itself, fewer than 5,000 years, some would say only about 2,000 years. Set in the context of either solar, geological, or biological time, then, and regardless of the grand narratives and myths that humans have constructed to glorify their importance, we are little more than *flickering instants of time*. We are, one might say, temporal contingencies in the grand march of time.

And yet, despite the reality that we are latecomers, we may very well turn out to be the ones who spoil the evolutionary party, at least in the local vicinity of space and time. For we are fundamentally disturbing deeper temporal orders. We are, to use the language of chaos theory, a strange attractor. We are precipitating disorder in Earth's organized physical, biological, and ecological systems. Perched on the cusp of the next millennium, and intensely conscious of the coming of the new millennium, we conceptualize time on scales that, however meaningful and reassuring to us, are radically incongruent with the temporal scales of evolution writ large. We shall consider four obvious ways in which time is out of joint.

9.2

First, conservation biologists tell us, our species has set in motion processes leading toward an anthropogenic mass extinction event.[7] Previous mass extinctions were the consequence of natural factors. The Cretaceous Extinction Event (some 65 million years ago) probably resulted from the earthly impact of a large extraterrestrial mass that generated a global encircling cloud of dust, which, by diminishing the amount of solar energy reaching Earth's surface, led to climate change. With climate change the age of the dinosaurs came to an end. The present event, now well underway, is the consequence of multiple human factors, not the least of which is our prodigious appetite for land and water, formerly available for other species, but now increasingly monopolized by ours alone. Within what is biophysically a virtual instant of time, that is, historical time, the human species is rending the fabric of life that evolved over the last 65 million years. Conservation biologists estimate that within a hundred years, if we continue on the present path, 50 percent of the known species on the planet will disappear, and that it will take 15 million years for biodiversity to recover.

Second, consider also that we are perturbing the atmosphere in a variety of ways, not the least of which is the depletion of stratospheric ozone.[8] This perturbation, given nature's temporal scales, is of almost instantaneous origin. It seems virtually impossible to believe that humans possess the power to upset a biophysical system that evolved over hundreds of millions of years. Yet, as Earth scientists point out, our activities are highly leveraged, meaning that the emission of even relatively small amounts of chlorofluorocarbons can have enormously consequential outcomes. Judged by the culturally dominant temporal frameworks, chlorofluorocarbons were technological and economic triumphs, demonstrating scientific mastery and enormous profits—in the short term, that is. Judged on the basis of nature's own scales, they have proved unmitigated failures, leading to the rapid depletion of the stratospheric ozone layer that protects life from harmful ultraviolet radiation.

A third manifestation of "temporal disjointedness" lies in the sheer numbers of humans now on this planet—a virtual population bomb, or explosion, of a single species.[9] We have become, within a single century, an ecological aberration, as Edward O. Wilson terms it. In all of biological evolution there has never been a species like our own—that is, a taker species, one that has vigorously and relentlessly captured and utilized as much of the material and energy flows on this planet as it can, and done so without consideration of physical, biological, and ecological consequences.[10] Paradox appears yet again, however, for babies are inherently good and of absolute value, so seemingly there can be no warrant for limiting the short-term reproductive success of the human species.

Fourth, and finally, humankind is caught in yet one more temporal predicament. We suffer from *chronic insecurity* itself. Over-determined by the temporal horizons offered by civilization, yet aware of the vaster, longer, and seemingly ageless cycles of nature, we suffer from temporal alienation. Ernest Becker's work, *The Denial of Death*, remains the best introduction to this topic.[11] Becker argues that our obsessive individualism and compulsive consumerism reflect anxieties about mortality. Even though the human species endures, our own lives seem short, and the death of our loved ones enormously and enduringly painful. We avert our attention from the finitude of existence through the projects of egoic consciousness and the pleasures of consumption. We live, thus, in a psychological condition of denial that the Earth matters.

9.3

In a wonderful book published some years ago, Neil Evernden argued that the human species is genetically predisposed to become alienated from Earth: hence, *natural aliens*.[12] Clearly this thesis cannot be readily dismissed. By all indications, Western culture is essentially oblivious to Earth matters, to the physical, biological, and ecological horizons within which life exists. Our dominant cultural vectors, those of political-economy, mainstream classical science and industrial technology, philosophy and religion, define a temporal trajectory *denying* that humankind is of and about the Earth, that Earth matters.

Evernden may be right, then, to the extent that he claims only that we have *become* natural aliens. What he overlooks is the fact human behavior is not entirely controlled but only conditioned by our genetic inheritance, which implies that we are not by nature "temporal aliens," but by culture. Which is to say that the human species is genetically under-determined and culturally over-determined.[13] This premise acknowledges the evolutionary history of our species while also affirming the role of cultural codes in human behavior. Our genetic constitution permits a plasticity of behavior. That is to say, our specifically human actions are mediated by cultural codes, codes (or memes) that can themselves be intentionally changed, at least within the limits of human nature itself.

Both fate and freedom are operating here. Both culture and nature can lead us to act in ways inimical to our well-being and that of the planet, and yet both offer an ambit of possibility, of freedom to change. By nature we are storytelling culture dwellers, or language animals, to use a different term.[14] Who would have dreamed that language would at one and the same time facilitate our survival, that is, the intelligent adaptation to the exigencies of existence, allowing us to emerge in the course of a few millennia as the overwhelmingly dominant species on the planet. And who would have dreamed that the same linguistic power would ultimately serve to sever and isolate us conceptually within our cultural cocoons from the deep past and chthonian energies of Earth itself.[15]

And so I reluctantly agree with Evernden that we have become natural aliens. But if we are, if indeed, time is out of joint, where does this leave us? What is the prognosis? And what is the recommended course of therapy?

Perhaps we should begin with our childhood.

9.4

The historical process by which we dissociated ourselves from nature begins with Neolithic villages and culminates in modern technological society. Entire books discuss this process in considerable detail, but space limitations preclude extensive discussion here.[16] Let me sketch in the picture with a few broad strokes of the brush. First, it is probable that the origins of Neolithic culture, of the cultivation of cereal grasses, domestication of herbivores, and permanent human settlement were rooted in a food crisis, itself caused by climate change accompanying the end of the last ice age, a climate change that led to the collapse of Pleistocene grassland ecosystems.[17] Second, it is probable that the evolution of cultural codes, especially those that mediate the economy and technology of survival, is driven by population pressure.[18] These first two factors, namely the prehistoric food crisis and population growth, historically reinforced each other, as permanent human settlement (sedentarism) and adequate nutrition countervail the infertility of women living in band society. The consequence is that once the ecological transition into the precursors of Western civilization has been made, there will be a continual and relentless evolution of cultural forms in response to the twin dynamic of increasing population and food production to sustain that population.[19]

Among these changing cultural forms were ideas concerning Earth and time. The Earth became less and less a sacred mystery, thought by humans to have existed long before their advent, and more and more was conceptualized as merely a physical or economic entity, mere property (a possession) to be technologically controlled and economically exploited. The emergence of the idea of history itself reinforced the devaluation of the Earth. History, of course, is virtually transparent to us, as we are so fundamentally embedded in "history" that its origins have become invisible, obscured behind the facade of contemporary life. The idea of history locates meaning almost exclusively within the scope of exclusively human experience. Thus, the Earth is only a mere stage upon which is enacted the human drama. Nature has no intrinsic temporal meaning; only when "men" bring nature inside the temporal process of history does it assume significance. History becomes a temporal cocoon into which humans retreat as they withdraw further and further from the green world of the deep past.[20]

Our own situation is remarkably akin and yet also dissimilar to that of our Neolithic forebears. Like them, we exist in circumstances of uncertainty, where climate change threatens our ability to feed a growing human population, and where the adaptation of culture to environmental exigencies is of great consequence. Unlike them, we have scientific insight and reflective awareness of the processes that have brought us to the present moment. We have rapidly increasing scientific understanding of the biophysical exigencies that confront us. And also unlike them, anthropogenic events have become major, even predominant, influences on planetary processes.

Given these factors, what is our prognosis? What is the likelihood that we might recover from our chronic insecurity; from conceptualizing the land as a commodity only, seeing it as nothing more than a resource for human appropriation? Or from believing that the atmosphere and oceans are nothing more than reservoirs for the byproducts of the industrial technologies that give us lives of comfort and convenience?

My prognosis is guarded. Pessimism about the future is clearly a self-fulfilling prophecy. Rosy-cheeked optimism is likewise self-defeating (and I must add that I put all plans for a sustainable future that are based primarily on technological fixes into this category). The wisest course for the immediate future is predicated on meliorism—that is, the notion that incremental changes in response to specific problems accumulate and interact over time, and that through this process cultural systems can be reconfigured. At some point in the future, historians will peer back into time and announce that a paradigm shift has occurred, that the old has been replaced by the new.

How then can we envision a course of cultural therapy that avoids the extremes of optimism and pessimism? We know that intentional change of cultural codes is always difficult but not impossible: The existing stories constitute both fate and freedom. We also know that the *Earth matters*, and that our dominant temporal narratives deny this reality. Thus, it seems clear that the possibility of adaptation to the exigencies of existence generally requires overcoming our alienation from the Earth. We can move in this direction by resolving the temporal incongruities between civilization and nature.

History offers ample evidence that the cultural codes which over-determine human behavior can be modified. To what degree these codes can be changed, how quickly, and by what means entail complex discussion beyond the present argument.[21] However, just a brief consideration might illustrate that we are *in media res*, that is, already involved in the process of incremental change. Consider three examples.

First, the neoclassical economic paradigm is undergoing delegitimation owing to temporal incongruities.[22] Quarterly profit-and-loss statements, yearly financial reports, natural resource policies predicated on benefit-cost analysis, and the obsession with the growth of national income accounts are now clearly understood as harmful to watersheds, to the atmosphere, to biodiversity and rain forests, and ultimately even to people. The biophysical anomalies of neoclassical economics are being addressed through new modes of analysis that incorporate more expansive temporal scales. The change in analytical form is quite simple to grasp. Consider the cod fishery off the coast of Newfoundland: If sustainably managed, the harvest of fish might have continued to provide jobs for thousands of people and food for millions. Literally, an unending stream of benefits. Because of pervasive mismanagement, a virtual primer in ignorance, the cod fishery has been destroyed.[23]

Second, the neoclassical political paradigm is also undergoing modification, even delegitimation.[24] For example, the "iron triangle" of public lands resource management has been broken, and new temporal scales are being employed in the policy-making process. (So-called iron triangles, constituted by commercial interests, federal agencies, and congressional appropriations subcommittees, dominated the policy process in a number of areas, such as forestry, ranching, and water resource development.) We have also recognized that the liberal democratic state does not exist only to protect property rights; wider commitments, such as the United States Wilderness Act and the Endangered Species Act, represent affirmations of cultural intentions and temporal scales that trump narrowly human economic interests and affirm geological, biological, and ecological values. And more radical notions of ecopolitics, beginning with Aldo Leopold's notion of the land community, are moving us toward a old-new conception of polity.[25]

Third, the dominant psychological theories that frame the human psyche as unique in evolution and as detached from biophysical processes are being delegitimated. Fields of inquiry, such as ecopsychology, paleoanthropology, sociobiology, and cognitive science, indicate that the lines drawn between humans and the rest of nature are not absolute boundaries but more permeable membranes, or matters of degree rather than kind.[26] The truth is that we humans were and remain primates, differing from our nearest kin because of a few extra wrinkles on the frontal lobes and a few other anatomical variations, especially in the vocal mechanism.[27] And we remain firmly embedded in nature. We deny these truths only at great peril to our being and to the rest of nature.

9.5

I come now to my conclusion apropos of philosophers and Earth scientists in dialogue on Earth matters. The three examples I have just mentioned should not make us overly sanguine that the human prospect is assured. Clearly, ideas that countervail the dominant narratives are appearing on the intellectual horizon. But the *dominant* narratives remain just that. Thus, the examples I have offered are not so much indices of a paradigm shift as a goad to further action on the part of ecophilosophers and Earth scientists. The philosophical tradition that assumes transcendental egos (Descartes's disembodied thinkers) can discover timeless and certain truths good for all people in all places must be overcome.[28] Similarly, the Earth sciences must overcome the naiveté that timeless and objective truths are revealed to passive scientific observers. Perhaps a new discourse might be created, one that we might call a "geophilosophical conversation." Such a conversation would be therapeutic, helping to heal both ourselves and Earth. The Earth does matter: Geophilosophical inquiry can help us limn its depths.

Geophilosophy might dare recontextualize species *Homo sapiens* as of and about Earth—as the very name Adam, in the sacred myth of Genesis, implies. Adamah: Earthman. Geophilosophical conversation might consciously deliberate the temporal scales used to frame, judge, and guide human (Earthman) behavior. Accordingly, geophilosophy might begin to reweave the discourse that currnetly defines an economics that maximizes profit while destroying biodiversity, a politics absolutizing human propertied interests while rendering the flora and fauna invisible, and a psychology isolating us from the biophysical matrix that created and still sustains us.

In truth, I would like to conclude on that note of affirmation. But the psychosocial reality is that the overwhelming majority of philosophers and Earth scientists are embedded in professional communities dedicated to the aversion of conceptual risk and the maintenance of the cultural status quo. In the twentieth century, philosophy has, with few exceptions, settled into a kind of comfortable complacency, cultural irrelevancy, and political impotency.[29] And twentieth century Earth science, at least viewed from my outsider's position, is dominated by the golden rule. Namely, those funding agencies that have the gold set the rules. Because those agencies themselves are funded through a highly politicized process with a truncated view of the cultural role of science, the potential of the Earth sciences, the very real potential inherent in the concept of deep time itself, to serve as a strange attractor within the cultural semiosis is not being actualized.[30]

Yet I for one try to resist such pessimism, which is unquestionably a self-fulfilling prophecy. The potential for a geophilosophical conversation is real. The opportunity cost for enjoining ourselves in that conversation is relatively small, particularly for those of us who have tenure. The outcome of a geophilosophical conversation could help us, as a species, overcome the temporal incongruities that lie at the roots of the malaise of the Earth.

NOTES

1. Ecological ethics, also called *ecoethics* or *environmental ethics*, is an interface discipline, a relatively new, multidisciplinary, interdisciplinary inquiry into an array of environmental dysfunctions, such as the extinction of species, depletion of stratospheric ozone, and water pollution. The term "interface disciplines" comes from Eugene P. Odum, *Ecology and Our Endangered Life Support Systems*, 2d ed. (Sunderland, MA: Sinauer, 1993).

2. Of course, things change. The collective epistemic community constituted by the Earth sciences is restive. Leaders within this community envision a greatly expanded, culturally vital role in the next century. Cf., e.g., Steve R. Bohlen, et al., *Geology for a Changing World: A Science Strategy for the Geologic Division of the U.S. Geological Survey, 2000–2010* (Washington, DC: U.S. Government Printing Office, 1998). In part, these changes are prompted by the failure of the traditional "social contract" that legitimated the expenditure of public resources on scientific research. See Radford Byerly, Jr., "U.S. Science in a Changing

Context: A Perspective," *Reviews of Geophysics, Supplement* (July 1995): A1–A16.

3. I will provisionally define a sustainable society as one that can produce the wherewithal necessary for its survival without undercutting the biophysical possibilities of future generations to do so. A truly sustainable society necessarily achieves economic sufficiency, maintains ecological integrity, uses appropriate technology, fosters human dignity, achieves social justice, and practices distributed decision making. Sustainability and programs for sustainable development (see Agenda 21, the U.N. plan for sustainable development in the 21st century, drawn up at the 1992 "Earth Summit" in Rio de Janiero) are not equivalent concepts.

4. Among many, see J. T. Fraser, ed., *The Voices of Time: A Cooperative Survey of Man's Views of Time as Expressed by the Sciences and Humanities*, 2nd ed. (Amherst: University of Massachusetts Press, 1981); and J. T. Fraser, *Time: The Familiar Stranger* (Amherst: University of Massachusetts Press, 1987). Fraser helps us realize that our "chronic insecurity" comes from perceived inconsistencies between, for example, psychological time as conceptualized by individuals in terms of a single life, and cosmic time, which seems to stretch to infinity.

5. The definitive study remains Nicholas Georgescu-Roegen, *The Entropy Law and the Economic Process* (Cambridge, MA: Harvard University Press, 1971).

6. Donald L. Baars, *Navajo Country: A Geology and Natural History of the Four Corners Region* (Albuquerque: University of New Mexico Press, 1995), 15.

7. Among many, see Edward O. Wilson, *The Diversity of Life* (Cambridge, MA: The Belknap Press of Harvard University Press, 1992); and R. Edward Grumbine and Michael Soulé, *Ghost Bears: Exploring the Biodiversity Crisis* (Washington, DC: Island Press, 1993).

8. Interestingly, the initial response to the scientific work documenting depletion of stratospheric ozone was denial. On depletion of stratospheric ozone, see John Firor, *The Changing Atmosphere: A Global Challenge* (New Haven and London: Yale University Press, 1990).

9. See, among many, Paul R. Ehrlich, A. H. Ehrlich, and J. P. Holdren, *Ecoscience: Population, Resources, Environment* (San Francisco: W. H. Freeman, 1977).

10. The processes of cultural evolution are not, of course, entirely understood. But population growth is perhaps the most consequential driver. See Allen W. Johnson and Timothy Earle, *The Evolution of Human Societies: From Foraging Group to Agrarian State* (Stanford, CA: Stanford University Press, 1987).

11. Ernest Becker, *The Denial of Death* (New York: The Free Press, 1973).

12. Neil Evernden, *The Natural Alien: Humankind and Environment* (Toronto: University of Toronto Press, 1985).

13. Evernden clearly acknowledges this point by affirming the premise that humans are secondarily nidicolous.

14. Among many, see Derek Bickerton, *Language and Species* (Chicago: University of Chicago Press, 1990); Charles Taylor, *Human Agency and Language: Philosophical Papers 1* (Cambridge: Cambridge University Press, 1985); and Will

Wright, *Wild Knowledge: Science, Language, and Social Life in a Fragile Environment* (Minneapolis: University of Minnesota Press, 1992).

15. *Ibid.* See especially Bickerton, *Language and Species*, and Wright, *Wild Knowledge.*

16. See Max Oelschlaeger, *The Idea of Wilderness: From Prehistory to the Age of Ecology* (New Haven: Yale University Press, 1991).

17. Mark Nathan Cohen, *The Food Crisis in Prehistory: Overpopulation and the Origins of Agriculture* (New Haven: Yale University Press, 1977).

18. See Johnson and Earle, *Evolution of Societies, supra* note 10.

19. I am glossing enormously complicated processes and histories of change. Among many relevant sources, see Clarence Glacken, *Traces on the Rhodian Shore: Nature and Culture in Western Thought from Ancient Times to the End of the Eighteenth Century* (Berkeley: University of California Press, 1967); and Paul Shepard, *Nature and Madness* (San Francisco: Sierra Club Books, 1982).

20. On the idea of history, see R. G. Collingwood, *The Idea of History* (London: Oxford University Press, 1956); and Shepard, *ibid., Nature and Madness.* On correctives for "human dis-placement," that is, the notion that Earth does not matter, that humans are isolated from nature, ensconced in the domain of history, see Carolyn Merchant, *Ecological Revolutions: Nature, Gender, and Science in New England* (Chapel Hill: University of North Carolina Press, 1989). She argues therein that "To see nature as active is to recognize its formative role over geologic and historical time. Only by according ecology a place in the narrative of history can nature and culture be seen as truly interactive" (p. 29).

21. Among many, see Bruce Lincoln, *Discourse and the Construction of Society: Comparative Studies of Myth, Ritual, and Classification* (New York: Oxford University Press, 1989); Wright, *Wild Knowledge*; and Taylor, *Human Agency, supra* note 14. Also see Max Oelschlaeger, "What Is Environmental Ethics?" forthcoming in John Howie, ed., *The Ley's Lectures* (vol. 4) (Carbondale: Southern Illinois University Press), for an interpretation of the discourses of environmental ethics in the context of environmental crisis and cultural adaptation.

22. See Georgescu-Roegen, *Entropy Law, supra* note 5, and Herman E. Daly, *Steady-State Economics: Second Edition with New Essays* (Washington, DC: Island Press, 1991).

23. Again I touch on complexities beyond the substantive scope and rhetorical slant of this paper. See, for example, Thomas R. McGuire, "The Last Northern Cod," *Journal of Political Ecology*, Vol. 4 (1997):41–54.

24. See, for example, William Ophuls and A. Stephen Boyan, Jr., *Ecology and the Politics of Scarcity Revisited* (New York: W. H. Freeman, 1992); and Roy Morrison, *Ecological Democracy* (Boston: South End Press, 1995).

25. See, for example, Daniel Kemmis, *Community and the Politics of Place* (Norman: University of Oklahoma Press, 1990).

26. For example, see Paul M. Churchland, *The Engine of Reason, the Seat of the Soul: A Philosophical Journey into the Brain* (Cambridge, MA: MIT Press, 1995).

27. See Philip Lieberman, *The Biology and Evolution of Language* (Cambridge, MA: Harvard University Press, 1984).

28. See Ilya Prigogine, *The End of Certainty: Time, Chaos, and the New Laws of Nature* (New York: The Free Press, 1997).

29. For a succinct account, see Chapter 3, "Re-visioning Philosophy and the Organization of Knowledges," in John J. Stuhr, *Genealogical Pragmatism: Philosophy, Experience, and Community* (Albany: SUNY Press, 1997).

30. On the cultural role of science, see among many Daniel Sarewitz, *Frontiers of Illusion: Science, Technology, and the Politics of Progress* (Philadelphia: Temple University Press, 1996).

1 0

A Sense of the Whole: Toward an Understanding of Acid Mine Drainage in the West

Robert Frodeman

Robert Frodeman teaches environmental philosophy and public policy at the University of Tennessee, Chattanooga. Possessing degrees in history, philosophy, and the Earth sciences, his work focuses upon the places where philosophic abstractions and everyday life intersect. He is currently completing a book on the philosophy of the Earth sciences, and is director of the Southwest Earth Studies Program, a National Science Foundation-funded educational experiment in philosophy and the Earth sciences held each summer in the mountains of southwest Colorado.

In the twentieth century, academic philosophy has distanced itself from the affairs of the world. In part, this has been the result of our culture's embrace of the assumptions of expertise—that the proper goal for all disciplines is the mastery of a specialized domain of knowledge. In contrast, Frodeman envisions the philosopher as the "professional amateur," helping our culture to articulate basic questions of origins and goals.

The environmental problems surrounding hard rock mining in the American West serve as a test case of this role for philosophy. Societal debates of issues such as the acid mine drainage controversy suffer from two problems. First, the cultivation of expertise has left experts unable to communicate effectively with either one another or with the general public. Philosophy can provide a powerful narrative framework for these various points of view. Second, our culture systematically misdefines political and philosophical questions as scientific or technical questions. Philosophy can expose this prejudice and thus demonstrate its relevance to the life of the community.

This essay examines the acid mine drainage controversy in the American West as a case study for understanding the role of philosophy in confronting our environmental problems. Environmental controversies are typically approached through a combination of science and public policy—science providing the facts necessary for decision making, and public policy representing the will of the people. This essay shows that a philosophical approach is needed if we are to adequately address the environmental questions we face today.

Philosophy has two roles to play in environmental controversies. First, it can provide an account of the generally philosophical (e.g., ethical, aesthetic, epistemological, metaphysical, and theological) aspects of environmental problems. This essay will argue that scientific data and the law are often used as stalking horses for concerns that are fundamentally philosophical in nature. Second, philosophy (and the humanities generally) can offer a narrative of the relations among the various disciplines (e.g., hydrology, chemistry, geology, public policy, and economics) that give insight into environmental problems. Such narratives can provide us with something that is sorely lacking today: a sense of the whole.

This essay, then, presents a different vision of the philosopher's (and by extension, the humanist's) role in society. Rather than primarily being a specialist who has mastered a particular subdiscipline of knowledge, philosophers are portrayed here as "professional amateurs"—people who use their training to help communities gain a sense of how the parts of our lives fit together. How does the truth that science discovers relate to our lived experience of the world? Or the hard facts of economics with the less tangible but no less real concerns about community values and quality of life? It is, of course, impossible to give a complete account of the various aspects of a problem such as acid mine drainage. Indeed, it has become clear that we are unable to provide a *complete* account of *any part* of controversies such as this one—although the continued narrowing of academic disciplines suggests that dreams of mastery through expertise live on. Thus, rather than a complete account, this essay offers what I will call a fractured narrative and an image of the acid mine drainage controversy.

As modest as this may sound, I believe it is sufficient. For in an information-rich age, when we can find massive, seemingly infinite amounts of knowledge on almost any subject, what is missing is such a grasp of the whole—in fact, precisely because of this glut of information. This was the original meaning of "logic": For the Greeks, *logos* did not mean the mechanical deduction of conclusions from premises, or the linking of cause and effect, but rather the sense of orientation and placement that comes from knowing how things fit together. Our society, then, while a monument to scientific and technical reasoning, has in this sense become deeply illogical.

Alasdair MacIntyre begins *After Virtue* with an account of the incoherent nature of contemporary ethical debates.[1] Using abortion as an example, MacIntyre recounts how different individuals, beginning from different sets of assumptions, reach irreconcilable positions. Our environmental conversations today possess a similar type of incoherence. This is true whether the controversy is the return of salmon in the Northwest, the burial of the nation's high-level radioactive waste at Yucca Mountain in Nevada, or the restoration of the Everglades ecosystem in Florida. Beginning from different places, individuals and groups marshal different sets of facts (or reject the relevance of facts), rely upon different kinds of experts (or deny the possibility of expertise), and appeal to different standards of evidence. Not surprisingly, such discussions often resemble ships passing in the night. Misunderstanding leads to endless debate, and frustration encourages the demonizing of one's adversaries. The political sphere, the place of public conversation, turns sour. People become alienated from the political process and view government as a distant and hostile entity. The result is public-policy paralysis, and finally litigation, as contestants turn to the courts to resolve their differences.

Nonetheless, progress toward the resolution of our environmental conflicts is still possible—if we can find the means for better communication between people coming from different interests and perspectives. But in claiming that the improvement of our environmental debates turns on bettering the means for communication, I am not suggesting that we should hire more experts in mediation and conflict resolution. Professionals in these areas have a role to play in the resolution of environmental issues; but the roots of our problems are to be found in our philosophic assumptions rather than in our techniques of mediation. We are the unconscious victims of epistemological assumptions that thwart conversation. Once we expose these assumptions we find that the conversation turns toward subjects that are seldom explicitly raised in environmental debates: questions of metaphysics, aesthetics, and theology.

It is our silence on these latter topics that renders so much conversation pointless or counterproductive. Our inability to integrate scientific and economic data and perspectives with our metaphysical, aesthetic, and theological concerns stymies serious discussion. And it is not only the claims of others that we misunderstand. We have also looked past our own deepest motivations, translating our concerns into language believed to be acceptable for public debate.

It will be noticed that the field of ethics is not emphasized in the list offered above. Questions of rights and obligations are of course pertinent to environmental debates. But the field of environmental philosophy has not been well served by its focus on ethics. Attempts to extend traditional ethical theories to apply to natural entities, whether those attempts are utilitarian, deontological, or aristotelian in nature, have not been particularly successful, either theoretically or practically.[2] This essay argues for a return to a larger landscape, where our ethical concerns about the environment are placed within a metaphysical, aesthetic, and theological context.

These points will be developed through the examination of questions surrounding acid mine drainage on abandoned mine lands in the San Juan Mountains of southwest Colorado. Acid mine drainage is a problem of national and global importance, but it has particular resonance in the western United States, where it is the greatest water-quality problem facing the region. The American West is home to as many as several hundred thousand abandoned mines, with thousands of miles of streams contaminated by low pH and high metal content. Sites in serious need of remediation may number in the thousands.[3] The question of restoring these areas—to what standard, and at whose cost—has sparked intense debate. Nationally, the size of the problem is enormous: One estimate puts the total cleanup at between \$32 billion and \$72 billion.[4] Parties to this dispute include landowners and local officials, environmental organizations and mining companies, lawyers, government scientists, tourists, and shopkeepers. The acid mine drainage debate is thus well suited for serving as a case study of the difficulties we face throughout our environmental controversies. The fundamentally interdisciplinary nature of the acid mine drainage controversy will also raise questions concerning the limits of knowledge and the role of expertise in resolving environmental disputes. Rather than being imposed from above, philosophy arises by attending to the interstices and points of contact between these various disciplines.

The acid mine drainage controversy also reveals the ontological nature of our environmental problems. Ontology is concerned with the ways that we divide reality up into parts, and the relationship among those parts. One of the first concerns of ontology is the question of "natural kinds." Are the divisions we perceive in the world written into the deep structure of reality, or are they to one degree or another expressions of particular historical circumstances, and thus subject to change as society changes? The disciplinary matrix of academia—the partition of the university into colleges, and colleges into the various departments—is itself a reflection of the ontological breaks we believe to have found in the world. We have broken knowledge into discrete disciplines—the arts and the sciences, poetry and geology, history and chemistry, aesthetics and political science—confident that these subjects may be studied in isolation from one another. Courses in art history are not required of the Earth science major, or vice versa. Neither is a course in Eastern religions included as part of the chemistry major. These decisions seem obvious today, but I believe that the inadequacy of our public conversations is rooted in precisely these choices. The current way of defining, and dividing, the disciplines has become a serious obstacle to resolving our environmental problems. But before we can turn to these ontological—and pedagogical—issues we must work through the details of our case study.

10.1 ACID MINE DRAINAGE IN THE SAN JUAN MOUNTAINS, COLORADO

By the standards of the well-watered eastern United States, the Animas River is not large. At the town of Durango, Colorado, winter flows average 500 cfs (cubic feet per second), and spring snowmelt may peak in late May at 7,000 cfs—numbers that translate into a shallow stream approximately 100 feet wide. (By way of comparison, the Tennessee River at Chattanooga averages 35,000 cfs, and reaches annual peaks of 160,000 cfs.)[5]

The Animas River's sources lie in the high reaches of the San Juan Mountains, a range of volcanic origin that rises to over 14,000 feet. Beginning as snowmelt, the Animas flows through Silverton and Durango, Colorado, and the Southern Ute Indian Reservation before joining the San Juan River in Farmington, New Mexico. Passing through the deserts of Utah, the San Juan River becomes part of the Colorado River. The Colorado, ponded by five major dams on its way to the Gulf of California, today seldom reaches the sea, its waters having been appropriated by cities and farms. The Animas itself is not a commercially navigable river[6]; its main use is for irrigation, fishing, and drinking water, and increasingly in recent years as a source of whitewater recreation.

The questions concerning acid mine drainage center upon the upper Animas drainage, in the high mountains and valleys surrounding the town of Silverton. The Animas River is formed by the confluence of three drainages: Mineral Creek, Cement Creek, and the upper Animas. The area is a popular tourist destination, attracting visitors for hiking, backpacking, horseback-riding, four-wheel driving, whitewater rafting, and the pleasures tied to the region's history. The latter include a narrow-gauge railroad, ghost towns, and historic mines from the gold-strike days of the Old West.

Southwestern Colorado has a rich history of mining.[7] The town of Silverton was center stage for a series of booms and busts that began in the mid-1870s. The current bust period dates from the closure of the Sunnyside Gold Mine in 1991. The closure of the Sunnyside was part of a shift in the U.S. mining industry—the end of the medium-sized operation, as mining, like agriculture and logging, became the domain of huge companies operating on the economics of scale. With the rise of new techniques such as cyanide heap leach mining it became possible to mine extraordinarily minute quantities of precious metals economically. In addition, a great deal of mining moved to foreign countries where the costs associated with land, labor, and environmental regulations are much lower.

Since 1871, over 9 million ounces of gold have been removed from the mountains surrounding Silverton, the second largest amount in the state (after the Cripple Creek district). [8] The miner's attentions were focused upon gold and silver, but the mountains also contain significant amounts of other metals such as zinc, copper, cadmium, lead, iron, and aluminum. When carried down into the streams, these metals have a variety of negative effects upon water quality. Zinc, copper, and cadmium dissolve in the water column and destroy aquatic life through their toxicity. Aluminum and iron precipitate on the stream bottom, disrupting the physical habitat of the benthic invertebrates (e.g., stone flies and caddis flies), which the fish depend upon, by filling in the pore space on the streambed that these insects use to breed.

The town of Silverton lies at 9,300 feet in a deep valley where the sun rises late and sets early much of the year. The winter population is about 350; access is gained via U.S. Highway 550, which is periodically closed in both directions by snowfall and avalanches. The population expands in the summer, with the return of absentee landowners (who come for the weather and the scenery) and the owners of small shops and galleries (who depend on the tourists). The town remains in the midst of the transition between mining town and tourist destination, and it has so far avoided being invaded by the chain restaurants, outlet stores, and mega-resorts that today characterize scenic mountain towns in the American West. Tourists come by car, or via the Durango and Silverton Narrow Gauge Railroad, which runs four times a day in the summer, taking three hours to cover the 46 miles from Durango.

The railroad first reached Silverton in 1882. Its presence allowed for easier transport of goods and ore, and led to the relocation of the main smelter to Durango the same year. Smelters in Silverton had not worked well because the high altitude makes it difficult for them to operate at high enough temperatures. The train has run continually ever since then, but beginning in the mid-1950s it has been devoted to mining the tourist trade. [9] The route from Durango generally follows the Animas, sometimes clinging to the sides of cliffs hanging over the river. This part of the trip, known as the "Highline," is not for the faint-hearted. The railroad also passes through the Weminuche Wilderness Area, the largest wilderness area in Colorado. Backpackers sometimes catch a ride into the wilderness on the train, to be dropped off at Elk Park or Needleton, to be retrieved later when they flag down the train on its return to Durango.

The veins and ore bodies today may have played out—at least until the next jump in the metals market. But the legacy of mining remains. A casual car-tour of the San Juan mountains is enough to alert one to the possibility of controversy. Several of the streams of the upper Animas (e.g., Mineral Creek, Cement Creek) run orange, with the water, rocks, and banks stained red and covered with an iron-aluminum sludge. A closer inspection reveals the absence of aquatic life, and raises concerns about possible health effects for the residents of Silverton, as well as for people downriver in Durango and Farmington. The region contains thousands of mining structures that still stand (or lean), and more than 1,500 abandoned mines. Depending on the viewer's aesthetic, ecological, and historical perspectives, these old mine structures, mine dumps, and tailing ponds are either picturesque, dangerous, or an eyesore. Today the entire town of Silverton is a National Historic Landmark, and the local economy survives on tourists interested in seeing a piece of western history and witnessing the natural beauty of the area.

10.2 DEFINING THE PROBLEM

Such, in outline at least, is the situation. To see the red and lifeless river courses of the upper Animas as a *problem* requires a shift in perspective. By what criterion does acid mine drainage count as a problem? What should count as a solution? Who are the responsible parties, and who should bear the costs of correcting this problem?

One finds no clear answers to such questions. The actions of recent years within the Animas drainage have not been triggered by any one particular law, individual, or group. Rather, like the river itself, events flow together, combining in often unpredictable ways. Issues of law and public policy play off individual personalities. Tradition and precedent pair off against scientific data. Supposedly objective scientific facts are found to be beholden to values and assumptions. And surrounding all of this are economic realities and the primordial human responses to a landscape that embodies the history and ideals of the people who inhabit and visit this place.

This multiplicity of perspectives means that there is no one way to describe adequately the story of acid mine drainage. Each account frames the story in a different way, highlighting certain features while casting others into shadow. The search for one right account is the search for a chimera: There is no single correct, objective perspective on the acid mine drainage controversy. One searches in vain for a framework or perspective that throws a clear and unequivocal light on all perspectives. What one has instead is a fractured narrative, which like a cubist painting offers a recognizable picture, but is made up of sets of angles, none of which dominates all the others.

This does not mean, however, that our only choice is to acknowledge that we are lost in the endless conflict of subjective opinions. The language of objectivity and subjectivity itself forms part of the problem. Human interests are always intimately intertwined with the production of knowledge, scientific or otherwise.[10] The most rigorously objective scientific procedure is always motivated by one or another set of personal or societal values, whether it be economic (generating profits or gaining tenure), political (improving community health), metaphysical (for the pure love of understanding the deep nature of things), or some combination thereof. Only when these various accounts are allowed to contest with one another is there the possibility of a more complete explanation emerging over time from the mosaic of perspectives. Community does not require conformity or complete agreement; rather, completeness is gained through the presence of ongoing contestation between different parties to a dispute.

Of course, in present-day American culture some perspectives are given priority over others. Claims based in economics, federal or state law, or science have a secure place within our cultural conversation, because they express our cultural assumptions about what constitutes real and objective knowledge. By contrast, ethical claims are more dubious. Although all of us lay claim to certain rights concerning freedom and justice, there is a deep suspicion that ethical claims are fundamentally arbitrary and self-interested in nature. Other human responses to the land—those that are aesthetic, metaphysical, or spiritual in nature—fare even worse.

In our case, that of acid mine drainage in the upper Animas River, the initial impetus for correcting the effects of mining came from enforcement of the federal Clean Water Act. But the application of the Clean Water Act was itself driven in part by the decision of a San Juan County commissioner and others to backpack fish into streams that they thought might be able to support them.[11] National and local environmental organizations did not raise the alarm about acid mine drainage; neither were there significant protests by the citizens of Durango and Silverton, the two towns most liable to be affected by poor water quality. In fact, local participation (in the form of the Animas River Stakeholders Group) was first organized by an outside (state) organization. Local citizens and officials were hostile to initial attempts to bring the two together, and overcame their reluctance to participate only out of fear that if they did not, state and federal agencies would simply dictate a solution from afar.

To understand the context of the controversies surrounding acid mine drainage, it is necessary to have a passing familiarity with some features of environmental law. These points of law provide the framework for the debate, and they determine much of what happens on the ground. Of course, interpretations of these laws vary; and officials have been known to search for statutory justification of what they want to do, rather than letting their actions simply be driven by the dictates of the law. It is also true that local citizens are often able to influence, obstruct, or ignore the implementation of these policies.

In 1972, Congress enacted the Water Pollution Control Act. This statute was supplemented by the Clean Water Act in 1977. Both acts, with subsequent amendments in 1981, 1987, and 1993, are commonly referred to as the Clean Water Act or CWA.[12] The CWA seeks to protect the nation's navigable[13] waters by setting water-quality standards for surface waters, and by limiting effluent discharges into these waters throughout the United States. The CWA's mandate calls for the "restoring and maintaining of the chemical, physical, and biological integrity of the nation's waters."[14] To achieve this, it requires states to establish water-quality standards for every river basin in the United States. The CWA also establishes a permit system to regulate point-source discharges (that is, pollution from discrete, identifiable sources such as a pipe or a ditch; nonpoint sources lack a single identifiable location, such as agricultural and city street runoff). Polluters are issued National Pollutant Discharge Elimination System (NPDES) permits that describe the stream-specific water-quality standards their effluents must meet.

The Environmental Protection Agency (EPA) has ultimate jurisdiction over enforcement of the Clean Water Act. But as a matter of practical policy, the EPA grants "primacy" over the application and enforcement of the CWA to state or tribal authorities that meet federal standards. The agency still retains an overall responsibility under the legislation, can approve or disapprove all state rules and regulations, and oversees enforcement. The EPA may also take independent action when it believes state programs are not adequately meeting federal standards. This represents a potential dilemma for a company or municipality: Having satisfied the state, the EPA may step in and take enforcement action. More commonly, however, the EPA arm-twists, threatening to take over a program until the state issues the rules and regulations the EPA wants. In at least some areas (e.g., EPA Region VIII, covering the Rocky Mountains) the EPA's exercise of authority has become increasingly subtle over time. The EPA strives to work with states on the front-end of rule-makings, rather than just rejecting them months or years later.

The 1972 water legislation called for the establishment of interim water quality goals throughout the nation by 1983. In response, the state of Colorado began moving toward control over the Animas drainage in 1979, when the Colorado Water Quality Control Commission (WQCC) first established use classifications and water standards for the river. At that time the WQCC did not attempt to classify the upper reaches of the drainage, because so much of the upper drainage was devoid of fish and macro-invertebrates as a result of low pH and heavy metal contamination. As noted above, this changed in 1985, when Bill Simon, then San Juan County commissioner, and others started a fish-stocking program.[15] Whether these streams contained fish prior to mining is a question difficult to answer, given the lack of historical information.

In 1991 the Colorado Water Quality Control Division (a different entity from the WQCC), part of the Colorado Department of Public Health and Environment (CDPHE), began to collect water quality data in the upper Animas, a program it continued through 1993. According to the Colorado Center for Environmental Management (CCEM), a nonprofit organization formed by Colorado Governor Roy Romer (1990-1998) to find solutions to environmental management problems, "this monitoring was prompted by a long-term need to better understand mine-related problems in the area and impacts across the Basin."[16] One might surmise as well that the hammer of the Clean Water Act had something to do with the monitoring.

In 1991, however, the slow grind of the bureaucratic machinery surrounding the Clean Water Act intersected with a second set of events. In that year the Sunnyside Gold Mine, the largest remaining gold mine in the San Juan Mountains, ceased operations. Echo Bay Mines, Inc., had bought the Sunnyside Mine in 1986; but after five years of losses Echo Bay shut down the operation. Gold production had never reached expected levels, and with the continued low price the company decided to cut its losses.[17]

As part of its operations, Sunnyside Gold Corp. (a subsidiary of Echo Bay) had a reclamation plan in place, but closure required that Sunnyside submit a final reclamation plan to the Colorado Division of Minerals and Geology. This plan called for removal of mining buildings, the consolidation and revegetation of waste rock and mine tailings, and the diversion of surface water flowing from the mine. The Colorado Division of Minerals and Geology approved the overall reclamation plan. Sunnyside nevertheless also needed a release from its NPDES permit from the CDPHE for the water that was leaving the site. The flow from the mine, which averaged 2,000 gallons per minute (gpm), was only mildly acidic, but it did contain high levels of zinc and iron. Prior to the CWA the water had flowed directly into Cement Creek. In the 1970s, however, the previous mining operator built a water treatment plant that operated under a NPDES water-discharge permit, the latter being issued by CDPHE. By the early 1990s, Sunnyside Gold was spending approximately $500,000 a year to run this plant, cleansing the mine water before it flowed into Cement Creek and eventually the Animas.[18]

As part of its mine closure plan, Sunnyside proposed to plug the mine entrance (known as the "American Tunnel") and discontinue treating the water coming from the mine. Sunnyside's claim was that the mine works would fill with water; soon the water within the mine would reach a chemical equilibrium similar to natural background conditions. The output from the mine portal would thus end, and any new springs that might appear in the area would have the pH and metal loading natural to the region. Sunnyside would achieve its goal—financial closure to its involvement at the site—with no negative effects upon the Animas drainage.

The CDPHE objected to the plan for two reasons. First, the treated water entering Cement Creek from the mine actually *improved* the water quality of the creek, which was affected by both natural and anthropogenic sources upstream of the mine. To plug the portal would therefore have the net effect of degrading the water quality of Cement Creek. Second, the CDPHE had doubts about Sunnyside's claim that the waters within the mine, once filled, would eventually equilibrate to natural background conditions. The natural conditions of the mountain had clearly been irretrievably changed: The mountain had been hollowed out, and fractured by continual dynamiting over the years. Furthermore, the production of acid drainage is greatly accelerated by exposing rocks to a mixture of water and air. Driving hundreds of adits and tunnels into the mountain had created the perfect combination of air and water for the production of acid drainage. Sunnyside Gold's plan was to keep the site entirely wet, thus turning off the production of excess acid drainage. The CDPHE, however, was far from sure that the flooded mine would equilibrate to natural background, doubts based upon complex geochemical and structural considerations.[19] Therefore, CDPHE refused to let Sunnyside out from under its water-discharge permit obligations, claiming that any new seeps that developed after the plugging of the American Tunnel would be subject to NPDES obligations as permitted discharges. Sunnyside's response was to sue in state court.

Sunnyside and the State of Colorado reached an out of court settlement in May 1996. As part of this agreement, Sunnyside signed a Consent Decree, which stated that it would clean up an "A" list of abandoned mined sites in the San Juans (only some of which belonged to Sunnyside). The goal here was the removal of zinc discharges approximately equal to the total amount of the discharge coming from the Sunnyside mine prior to treatment. While zinc is not the only bad actor among the metals in the Animas drainage, it was chosen as a standard because of its "conservative" nature. In contrast to other metals, zinc stays in the water column, thus giving a better overall sense of improvements in water quality.

Monitoring of zinc levels at a site known as A72 (on the Animas just downriver of Silverton) would continue for five years after this cleanup was completed. If zinc loading remained at or below an agreed upon baseline level (approximately 550 ppm) over this period, the expectation was that Sunnyside would be released from its permit, and could turn off its water treatment plant and walk away (or perhaps turn the treatment plant over to another entity). The long-term goal for the stretch of river from Silverton to Elk Park (known as "4a") was a zinc load of 225 ppm. It was thought that such a reduction was sufficient for the river to support a population of brook trout. If for some reason conditions did not improve, Sunnyside had a "B" list of sites that could also be remediated.

By 1999 all of the orphaned sites on the "A" list had been remediated—at a cost to Sunnyside of approximately $28 million. To date, monitoring at A72 has not shown any improvement in zinc levels. This is despite the fact that Sunnyside continues to run its water-treatment plant. This has understandably left officials at Sunnyside Gold perplexed and searching for explanations.[20]

We shall return to the issues raised by this Consent Decree. First, however, note how the terms of the debate shifted. The reasonableness of the original Sunnyside proposal now turns not only on the details and the interpretation of the Clean Water Act, but also on our understanding of the chemistry, hydrology, and geology of the region. Here two disciplines meet—public policy and the Earth sciences. Let us, then, turn from the CWA to the situation on the ground—and beneath it.

10.3 SCIENCE AS HERMENEUTICS

Approaching the San Juan Mountains from the south, driving up the Animas River valley past the towns of Durango and Hermosa, one passes beneath massive cliffs of red and buff-colored sedimentary strata that tilt downward to the south. Coming to the San Juan Mountains from the north past Ridgway and Ouray, Colorado, one faces a similar scene, but now the layers slant downward to the north. It is as if a titanic force had pushed the sedimentary beds from below, tilting the beds until they burst through at what is now the center of the San Juan Mountains.

This is approximately the account that geologists offer of the San Juan Mountains. The sloping sedimentary strata all point toward the center of the mountains, where one finds abundant evidence of volcanic activity: lava, welded ash flows, mineralization, and deeply faulted structures. Sometime in the mid-Tertiary—about 35 million years ago—what is now southwest Colorado was a volcanic landscape. The source of the lava and ash was probably a batholith, a huge mass of subterranean magma that was also the origin of the minerals that would later interest the miners. The volcanism continued over 10 million years, consisting of a long series of eruptions, many of which dwarfed Mount St. Helens and Mt. Pinatubo. Much later—beginning approximately 2 million years ago—the mountains were covered and sculpted by glaciers, giving the valleys the distinctive U-shape that they possess today. The glaciers have advanced and retreated here as elsewhere, the last retreat occurring 15,000 years ago.

Silverton itself lies in the midst of the San Juan volcanic field, at the edge of the Silverton caldera (Figure 10.1). A caldera is a volcano that has collapsed upon itself: After the explosive venting of the lava and ash the volcano collapses, leaving a concave depression. Although interpretation of the area is complex, there is evidence of multiple periods of volcanic activity, and multiple, overlapping calderas in the region. One estimate puts the number at 15. It is clear, however, that the hydrothermal activity related to the collapsed volcanic complex is the source of the heavy metals in the region. One finds a series of faults running along what is interpreted as the rim of the caldera, and another set of radial faults that issue from what appears to be the center of the volcano. These fractures later served as the plumbing system for the upward movement of mineral-laden fluids, which precipitated metallic ores in veins along these fractures.

Seeing the caldera requires an educated eye. The volcanic lavas and welded tuffs (superheated volcanic ash that falls like snow and adheres) are clear enough. So are the signs of mineralization in the area: Red Mountain north of Silverton gets its name from the oxidized orange and red stains covering its sides. But the geography has

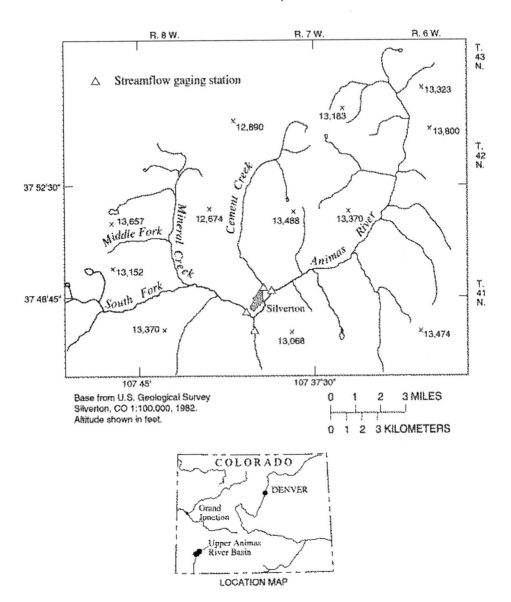

Figure 10.1   Upper Animas River Drainage & Silverton Caldera

been transformed since the mid-Tertiary. The topography has become inverted: What was once high is now low, and vice versa. The faulted outside edge of the caldera defined an area of weakness, which not only allowed the passage of hydrothermal fluids from below, but also snowmelt and rainwater from above. Erosion attacked the ring faults, especially during the last 2 million years of glacial conditions in the area. Thousands of feet of rock were removed, and the fault zones came to define the paths of the river courses. Glacial valleys were cut along the edge of the caldera at the same time as the area was elevated through regional uplift. Today the river courses of

Mineral Creek and the Animas River define the south and west sides of the Silverton caldera (Figure 10.1).

One point to draw from this—beyond the intrinsic interest of such a marvelous series of events, in a landscape shaped by both fire and ice—is that the region is a naturally mineralized area subject to acid drainage and heavy metal contamination long before the appearance of humans. After all, it was these naturally occurring conditions that drew the miners to the region in the first place. Acid *mine* drainage, then, is but an accelerated form of the natural processes of the acid *rock* drainage intrinsic to the area. Acid rock drainage results from natural weathering processes, biologic activity, and the regional geology.

However, several hundred adits and tunnels have been driven into the rock of the Silverton caldera and surrounding area. The mountains have been further fracturing through the use of dynamite. Finally, by dumping mine tailings and waste out on the surface, the overall area exposed to air and water has been greatly increased. In such circumstances, separating natural background conditions from what was caused by human activity becomes a difficult and contentious question.

The production of acid drainage is a complex process involving chemistry and biology as well as geology. Exposure of the sulfur-rich rocks to air and water causes the sulfide minerals (e.g., pyrite, galena, and sphalerite) to oxidize. Take the example of pyrite (FeS_2), commonly known as "fool's gold." Rainwater, snowmelt, and exposure to air break the iron sulfate compound into its constituent parts, ferrous iron and sulfur, at a relatively slow rate. The sulfate ions react with the water to produce sulfuric acid, and the iron goes into the water as well. However, this chemical reaction by itself is not sufficiently energetic to produce significant amounts of acid drainage, which does not occur until the reaction is massively accelerated by the biological activity of sulfur-oxidizing bacteria from the genus *Thiobacilli*. *Thiobacilli* are chemolithotropes, which "eat" rocks, acquiring energy from the minerals. *Thiobacilli* rapidly increases the rate of sulfide oxidation, resulting in an explosive expansion in the amount of acid drainage.

The pH value of a solution is a measure of acidity based upon a logarithmic scale: Going down the scale, each number represents a tenfold increase in the amount of acidity. Thus, the difference between a pH of 7 and 3 is four orders of magnitude. The pH levels in some of the headwater streams of the upper Animas are as low as 2 and 3 units (lower than the pH of vinegar; trout die at pH values below 5.4). A low pH also allows the heavy metals to stay in solution, causing the high concentrations of metals in the water that leads to the fish kills. If the pH is raised (for instance, by passing over some limestone), some of the metals will precipitate out of solution, leaving the bright orange-red stains coating the rocks on the sides and bottom of the stream. As noted earlier, this destroys the habitats for the bugs that the fish feed upon.

Distinguishing between natural and anthropogenic acid drainage is a tricky process. Of course, an old mine with timbers askew and a thick rivulet of red gunk issuing from the portal is a poster child for acid mine drainage runoff. But it remains an open question how much of the discharge has been generated through mining, and how much is simply the concentration in one location of natural runoff that previously found its way to the surface through unknown springs across the mountain. The geologist offers an educated guess: In this case, it would probably be that the majority of the drainage is human-caused. But conditions are sufficiently open to differing interpreta-tions to give rise to interminable debates. Indeed, in the mining industry one com-monly speaks not of acid mine drainage (AMD) but of acid rock drainage (ARD).

Hydrologists at the U.S. Geological Survey have worked extensively to distin-guish natural from anthropogenic acid drainage. One part of the Animas watershed, the Middle Fork of Mineral Creek (Figure 10.1), was found to contain 73 springs and 17 mines.[21] Of the mine sites, seven had water coming from the portal. Throughout the basin, hydrologists face the challenge of trying to identify whether seemingly "natural" seeps were truly natural, or whether they are the surface expression of mining activity further upslope. Sometimes the evidence is conclusive; but often hydrologists are faced with trying to map the faults in an area to help define the possible relation between an old mine site and a seemingly natural spring.

The fundamentally interpretive nature of such phenomena is visible at the Red Chemotroph Spring in the Cement Creek drainage. A group of students and I were led to the site by a USGS hydrologist who in subsequent publications offered this as an unambiguous example of a natural mineral-rich seep. As our group listened to the description of this as a natural spring, I wandered upslope and found a prospect pit and two cables snaking down the mountain. Returning to the group, I asked our guide about these signs of human activity, and whether Red Chemotroph Spring might not be anthropogenic in nature. Our hydrologist stated that he remained confident that Red Chemotroph was natural in origin.

There was no reason to doubt his judgment; this was a hydrologist who had spent several years studying the region. In fact, the life of the field scientist is filled with situations such as this—judgment calls made on the basis of one's education, par-tial data, and years spent in the field. Most of the evidence surrounding acid mine drainage involves such judgment calls. For instance, the pH and mineral content of a stream can vary with both time of day and season. On a warm summer day a small rivulet will be flushed with snowmelt by noon, a fact that will itself vary according to whether the winter was one of light or heavy snowfall. Readings in early June will dif-fer from July or September as the snowpack declines. Readings by two people on the same day, at the same time, in two valleys, can still vary because of different surface or weather conditions.

Even the instruments measuring pH and conductivity can be thrown off by the lack of ions present in the snowmelt. Measuring pH—the amount of protons in a solu-tion—requires a basic level of background conductivity. But rain and snow are typi-cally low in anions and cations. Unless the fieldworker adds potassium chloride to the sample the result will be systematically in error. Correcting for conditions such as these requires a nuanced sense of one's subject matter, what the biologist Michael Polanyi has called the "tacit dimension of knowledge." Aristotle noted the existence of a simi-lar intellectual skill when he spoke of the central role of *phronesis*, or judgment, in thinking.[22] Judgment depends upon the ability (and the opportunity, increasingly rare in a fast-paced culture) to deliberate rather than calculate. The exercise of judgment requires a nuanced appreciation of the details of a situation that cannot be reduced to a set of rules.

All of these issues are present in the debate between Sunnyside Gold Corp. and the State of Colorado. Take, for instance, the question of why the remediation of the

"A" list of sites in the basin has not led to improved water quality at monitoring site A72 below Silverton (Figure 10.1). There are any number of possible explanations for this: The period used as a baseline for judging water quality could be aberrant, reflecting unusual climatic variability during the initial period of monitoring. Or the current period may be unusual, thereby skewing the readings. It is also possible that the system has not yet responded to the changes—it could take years or decades for the effects of the cleanup to register at A72.

When such issues are raised, field geologists often apologize about the fundamentally interpretive nature of their research. After all, "professional judgment" sounds suspiciously like "subjectivity." Such an appeal flies in the face of our culture's image of science, which is supposed to offer a precise and certain mathematical basis for policy decisions. Field sciences such as botany, ecology, hydrology, and geology are thus typically seen as poor kin to laboratory sciences, which promise reliable (that is, repeatable) results. What goes unappreciated is the fact that such "lab results" are themselves unreal—the variability that the field scientist confronts is the variability of the real world. Field scientists have developed their skills at making sense of the hints contained in the rocks or the water. Hermeneutics, or interpretation theory, is a type of reasoning that relies as much upon experience and discernment as it does calculative ability.

The question of the nature of "field reasoning," and its supposed inferiority to the laboratory sciences, is a topic that deserves more attention. The point to be emphasized here is that science is typically brought into political controversies because it is seen as the means for resolving debates. The conflict between Sunnyside Gold and the State of Colorado was going to be mediated on the basis of sound science. Instead, the science itself became a bone of considerable contention.

There is one final element of the acid mine drainage controversy that requires exploration. It is arguable that this entire argument has gone off track. What difference does it make whether the streams of the upper Animas are stressed as a result of mining, or are "polluted" by naturally occurring springs and seeps? A pH of 3.2 is a pH of 3.2 in any case. This question is seldom faced head-on, but its presence hovers about the topic like swamp gas. To address it takes us beyond the discourses of politics and science to subjects that are rarely taken seriously in our public environmental debates.

10.4 THE METAPHYSICS OF ACID MINE DRAINAGE

Assume for the moment that the scientific research done to date is correct: that a great deal—quite possibly more than half—of the acid drainage and heavy metal contamination in the rivers is the result of natural geologic conditions, thus predating any mining activity. Of course, identifying which streams are "naturally polluted" is a highly interpretive exercise. But set this question to one side, and consider: If a stream is found to be naturally lifeless, do we leave it be, and clean only those areas that have been rendered sterile through human action? What if we discover that a human-acidified stream is much more expensive or harder to clean (because of accessibility, or local geology) than an equally "polluted" naturally acidified stream? By what reasoning would we spend the extra money to clean up the human-caused damage?

These are peculiar questions, leading to points that are abstract and even metaphysical in nature. But they are also immediate and practical, in that they express a concern and an intuition deeply felt by many. Followed out, such questions draw us into provinces seldom seriously explored within contemporary political debates. They ask us to consider carefully our motivations for cleansing these streams. Are we doing it because of human health and safety (i.e., questions of water quality, to protect our drinking water)? Are we doing it to create habitats for fish and invertebrates, even if they were not native to the area? Are we doing it to increase the tourist trade through the expansion of trout fisheries? Or—as was once suggested in my presence—because the tailings piles and stream courses "look like sin"?

Such questions approach us from two sides. First, as we have seen, the appeal to science as a sufficient basis for resolving environmental problems fails. Science helps us draw parameters around a problem, or outline possible scenarios for action. But science only rarely can substitute for political deliberation. More commonly, science provides us with what Daniel Sarewitz calls an "excess of objectivity." Science is a sufficiently rich and diverse enterprise as to provide support for a variety of reasonable and often contradictory positions on a subject.[23]

Second, our responses to nature cannot be encompassed by the calculus of science and economics. A lifeless, foul-smelling river provokes our disgust, and so we turn to the chemist and the public health official to define our outrage. The cutting down of old growth forests has a severe impact upon fisheries downstream, and we ask the ecologist and economist to quantify our losses. But the paucity of this approach becomes apparent once we examine our own reactions to environmental destruction. We are sickened by a fouled stream even if our water supply is located elsewhere; and the conversion of an ancient forest into a redwood deck evokes in many a sense of sacrilege.

In the upper Animas, concerns with the impact of tailings piles and polluted rivers on human health have played only a minor role in these debates. The town of Silverton draws its drinking water from the Boulder Creek watershed, which was withdrawn from mineral entry in the nineteenth century. By the time the Animas River reaches larger towns downstream, such as Durango and Farmington, the flow has been sufficiently diluted so that metals are not a concern.[24] The small amount of epidemiological research conducted in the Silverton area has not suggested a serious problem.[25] Finally, if the primary motivation for cleaning up the area is economic (either through improving the scenery or the fishing), then it would be much more cost-effective simply to write the citizens of Silverton a check.

No, something more obscure and fundamental is going on here. There is on the part of many an intuition that there has been something wrong with the way we have behaved. We have mistreated the natural world, and we are under some type of obligation to correct our mistakes.

Is it possible to make sense of this intuition? We are all aware of the criticisms leveled at such claims. They are rejected as subjective and unnecessarily mystifying. Moreover, they are often seen as politically dishonest—matters of personal preference masquerading as something more, laying claim to an illegitimate place within political discourse. For these reasons, such intuitions have no standing in public debate. In brief, they are ejected on epistemological grounds: They do not pass the reality test of being "real" or objective.

This response, however, presupposes the existence of a higher standard of certainty and proof. It is the contrast with the purported clarity and objectivity of scientific knowledge that renders these other claims to knowledge and reality otiose.

The status of scientific knowledge thus becomes central to the evaluation of the acid mine drainage controversy. In recent years this question has been the subject of a contentious debate—the "science wars"—where the defenders of the objectivity of the scientific method battle those who seek to demonstrate that science is just as subjective as politics, aesthetics, and religion. To pursue this debate would require another essay, but a reasonable case can be made that science is less than wholly objective, and fields such as aesthetics and metaphysics are less subjective, than has been assumed. The type of reasoning practiced by the *field* (rather than laboratory) *sciences* provides us with a more realistic model for the nature and limits of human reasoning, both scientific and otherwise.[26]

But let us return to the intuition described above: Is it possible that our responses to the land are fundamentally shaped by metaphysical, aesthetic, or spiritual commitments? One indication of this is found in the 1964 Wilderness Act, which defined wilderness as "an area where the earth and its community of life are untrammeled by man, where man himself is a visitor who does not remain." Wilderness is identified as an area which "(1) generally appears to have been affected primarily by the forces of nature, with the imprint of man's work substantially unnoticeable; [and] (2) has outstanding opportunities for solitude or a primitive and unconfined type of recreation."[27] However one parses out such statements, they point toward a nonutilitarian view of parts of the natural world. A second hint is found in the right of Native Americans to protect their sacred sites, given legal status under the provisions of the American Indian Religious Freedom Act of 1978 (AIRFA), which guarantees Native Americans right of access to these sacred areas.[28]

To cast light on these claims, consider Robert Elliot's argument in the essay "Faking Nature," which explores what he calls the ontological aspects of our concern with nature.[29] What, if anything, is at stake if a mining company can completely restore an area after the mining is completed, to the point that no one could tell the difference afterward?[30] Of course, many will doubt the premise—that it is possible to put an ecosystem or the natural features of the land back together again. But to argue on these grounds is to make the preservation of nature dependent upon technological insufficiency. If (or as many will claim, *when*, given the rapidly evolving nature of technology) we are able to reconstruct ecosystems, this objection to the plans of the mining company becomes irrelevant. One sees the same type of argument when it is suggested that the rain forest should be preserved because of the medicines that have been developed from tropical plants, or because of the not-reproducible ecosystem services that the forest provides. These arguments also evaporate when the medicines are able to be synthesized in the lab, or when we improve our understanding and technological prowess in reconstructing or mimicking ecosystem services.

The virtue of Elliot's argument is that it highlights the fact that our concerns are often at root metaphysical or ontological in nature. Elliot offers a series of examples that show that it is often the origin and ontological status of a thing, rather than its particular physical constitution, that matters to us most. Imagine a beautiful, hand-crafted knife received as a gift. One cherishes the knife, and displays it prominently in one's home. It is then discovered that the gift was made from the bone of a person killed expressly for the purpose of making the knife. Nothing has changed in terms of the knife's chemical or physical characteristics. Nevertheless, its nature has been irretrievably changed. The knife is now a source of revulsion rather than delight.

Consider another example. Stand in the valley carved by the South Fork of Mineral Creek west of Silverton, and look east to the massive shape of Anvil Mountain jutting into the sky. The peak makes a stunning impression, not only through its height and massive shape, but also because its peak and rocky slopes are an expanse of yellows, oranges, and reds that stand in brilliant contrast with the evergreens covering its flanks. But what if it is discovered that the colors come, not from natural processes of erosion, but from the drainage from old mines at the mountain crest? We find that the beauty of the mountain is related to its ontological status, because the *meaning* of the colors changes for us depending upon whether they were produced through natural processes or are the result of mining.

Of course, such questions immediately take us into debates on the status of the aesthetic object. And some would argue that the origins of an object are irrelevant to evaluating its formal qualities as an aesthetic object. But no matter where one finds oneself in this debate, aesthetic questions, which have been long viewed as having no practical import, have now become crucial to the question of whether to repair the damage of acid mine drainage.

Robert Elliot notes that to acknowledge the value of natural things does not commit oneself to affirming *all* natural things. One may grant that sickness and disease are natural while still combating them. Neither does the point require an absolutely pristine sense of the natural. One may argue that things are more or less natural: Indeed, today it is doubtful if there is anything in nature that has not been modified by human activity in one way or another. Furthermore, one can object to Elliot's assumption that the natural is something that once lost is gone forever. As Irene Klaver has argued, human-modified landscapes can regain their naturalness by being allowed to once again go their own way.[31] Elliot's point is simply that, within certain limits, the naturalness of a thing provides us with a reason for protecting it. Or to reframe the point, the *meta*-physical status of an object counts as well as do its physical characteristics.

The debate over acid mine drainage has a theological or spiritual cast as well. A mountainside strewn with rusting pipes and corrugated steel and torn up by bulldozers evokes in many a sense of violation. The historian of religions Mircea Eliade has argued that it is this sense of trespassing a limit that lies at the root of the religious sensibility.[32] The sacred identifies something more elemental and intuitive, and less organized or doctrinal, than what falls under the headings of "religion" or "church." Rather, it expresses the sense that there should be limits to human behavior, and without such limits humans lack an orientation in life.

Still other aspects of the acid mine drainage controversy drive us toward philosophical puzzles. Consider Sunnyside Gold's desire to be relieved from its financial obligations under its NPDES permit. Sunnyside understandably wants to limit its liability for future claims against the corporation. But estimating the possibilities for further damage from its mine workings presents us with epistemological quandaries, as well as ethical and speculative questions concerning our relations to the future.

Assume for the moment that zinc levels at the monitoring site A72 decline after the remediation of the orphaned sites in the area. How could one possibly demonstrate that this is a causal relation, rather than a correlation based on other factors? If Sunnyside was released from its permit obligations, what assurance could we have that zinc levels will not spike in five or 50 years? One USGS researcher estimates the time frame for the San Juan ecosystem to return to natural background conditions at between 2,000 and 10,000 years—the latter period longer than recorded human history.[33] The effects of mining will echo across the centuries; but how do we fairly pass out responsibilities that will outlast our civilization? There are no clear or easy answers to such questions. But neither are they avoidable.

10.5 CONCLUSIONS

So we arrive at the portals of philosophy. The point here has not been to provide answers to the philosophic riddles surrounding acid mine drainage, but rather to demonstrate the inevitability of our having to confront these riddles. Questions like the ones identified require that we embrace perspectives that our culture has treated as superfluous. For it turns out that traditional questions of metaphysics, theology, aesthetics, and politics—the nature of the good life, the sacred, the beautiful, and our obligations to each other and to nature—are as intrinsic to public debates as are the much more recognized disciplines of science, politics, and economics.

The discounting of humanist perspectives has occurred through a combination of scientism (the belief that science provides a uniquely objective means to knowledge) and the dominance of a libertarian political philosophy. Libertarianism claims that we should privatize the question of the good life, bracketing public discussion of broadly philosophic questions in order to allow each individual to answer these questions in his or her own way. Libertarianism does so in part because of its scientistic assumption that questions concerning the good life are irredeemably subjective and thus not susceptible to rational analysis. But libertarianism is also based upon another assumption: the existence of an endless frontier. The formative American experience of the frontier—the presence of a seemingly unbounded expanse of space and resources—meant that there was little need to debate difficult philosophical questions. In the case of conflict one could simply move on to another place. In contrast, issues such as acid mine drainage point up that there is no longer an "elsewhere" to move when we degrade the environment.

But the need for an interdisciplinary approach to our environmental problems involves something more than simply combining the perspectives of the scientist and the politician with those of the humanist. In part, our challenge is to sharpen our sense of when the questions we ask leave one genre for another—when a political question has turned into an economic or spiritual question, or a scientific question into an epistemological or metaphysical one. That is, we need to acquire greater skill at ontology. But if there is to be effective communication between the scientist and the historian, or the public policy analyst and the aesthetician, we must also develop a common body of vocabulary and experience.

Thus, in addition to training a class of experts, our environmental challenges require "specialists of the general," individuals whose task is to acquire at least the image of the whole. Such broadly trained academics would serve as a counterweight to the dominance of narrowly trained experts. Where might we find such academics? Creating broadly trained individuals has been the traditional goal of the liberal arts education. But this goal has been abandoned by doctoral programs in the humanities, which have instead defined a graduate humanities education in terms of specialization and expertise. Reorienting at least part of graduate study in the humanities toward the training of "specialists of the general" would not only improve the quality of our environmental debates but could also lead to the revitalization of the humanities, by returning the humanist to public life.[34]

ACKNOWLEDGMENTS

Thanks are due to the Natural Resources Law Center at the University of Colorado and the El Paso Energy Foundation, which supported me as the El Paso Energy Fellow during the research and writing of this essay. The following individuals have answered questions or made suggestions concerning the acid mine drainage controversy: Carol Russell, EPA Region VIII; Kirk Nordstrom, U.S. Geological Survey; Christopher Hayes, Ireland Stapleton Law Firm; Larry Perino, Sunnyside Gold Corp.; Douglas S. Kenney, Natural Resources Law Center; Robert Blair, Fort Lewis College; Bill Simon, Coordinator, Animas River Stakeholders Group; and Irene Klaver, University of North Texas. I also thank my research assistants, Derek Matthiessen, Peter Nichols, and Andy Vogt, for their help.

NOTES

1. Alasdair MacIntyre, *After Virtue: A Study in Moral Theory* (South Bend, IN: Notre Dame University Press), 1982.
2. This point has been made by Bruce V. Foltz, "After Heidegger and Deep Ecology: Toward an Environmental Aesthetics," forthcoming in a Northwestern University Press collection of essays edited by James E. Swearingen and Joanne Cutting-Gray.
3. Cf. The highest estimate of several hundred thousand mines in the West comes from the Mineral Policy Center, which can be found in its book, *Golden Dreams, Poisoned Streams* (Washington, DC: Mineral Policy Center), 1997, p. 13.
4. *Ibid.*
5. Water data available at http://waterdata.usgs.gov.
6. "Navigable" is here meant in the common, rather than the technical sense of the Clean Water Act (see below).
7. On Colorado mining history, see Duane A. Smith, *Mining America: The Industry and the Environment, 1800–1980* (Lawrence University Press of Kansas), 1987. A good general introduction to the San Juans is *The Western San Juan Mountains: Their Geology, Ecology, and Human History*, Robert Blair, ed. (Niwot: University of Colorado Press), 1996. For Durango, see Duane A. Smith, *Rocky Mountain Boom Town: A History of Durango, Colorado* (Boulder: University of Colorado Press), 1992.

8. Interview with local historian Bill Jones, Montrose, Colorado, April 14, 1999.

9. Interview with Bill Simon, Animas River Stakeholders Group Coordinator, April 23, 1999; Beverly Rich, Chair, San Juan Historical Society, April 14, 1999.

10. See Jurgen Habermas, *Knowledge and Human Interests* (Boston: Beacon Press), 1967, for a classic statement of this position.

11. Bill Simon, E-mail correspondence, January 28, 1999.

12. For a general overview of this legislation, see Olga L. Moya and Andrew L. Fono, *Federal Environmental Law: The User's Guide* (St. Paul, MN: West Publishing), 1997. p. 288.

13. Under the Clean Water Act, "navigable" is defined as including almost any body of water, with the exception of groundwater, including potholes, intermittent streams, dry washes, canals, and wetlands.

14. *Federal Environmental Law: The User's Guide, supra* note 12, p. 289; Clean Water Act §101, 33 U.S.C. §1251 (1972).

15. *Supra* note 11.

16. *Animas Status Report*, Colorado Center for Environmental Management, May 1996.

17. Interviews with Chris Hayes, attorney for Sunnyside Gold Corp., February 11, 1999, and Carol Russell, EPA, Region VIII, February 16, 1999.

18. *Ibid.*, interview with Chris Hayes.

19. These include the possibility that efflorescent salts that formed during mining could lead to the continuing production of acidic waters even after flooding, and the difficulty of completely flooding the mine due to myriad fractures and seasonal variations. Personal communication, Kirk Nordstrom, U.S. Geological Survey, April 15, 1999.

20. Interview with Larry Perino, Sunnyside Gold Corp., April 12, 1999.

21. W.G. Wright, "Natural and Mining-Related Sources of Dissolved Minerals During Low Flow in the Upper Animas River Basin," Southwestern Colorado, USGS Fact Sheet FS 148–97, October 1997.

22. Michael Polanyi, *Personal Knowledge: Towards a Post-Critical Philosophy* (New York: Harper & Row), 1964; Aristotle, *Nicomachean Ethics*, Book 6.

23. Daniel Sarewitz, "Science and Environmental Policy: An Excess of Objectivity," in this volume.

24. Although in 1906 the city of Durango attempted to sue the Gold King Mine and San Juan County for contaminating their water source, today its drinking water comes from another watershed, that of the Florida River.

25. Personal communications with Durango District office, Colorado Water Quality Control Division, Colorado Division of Public Health and Environment, April 2, 1999.

26. Cf. Robert Frodeman, "Geological Reasoning: Geology as an Interpretive and Historical Science" *Geological Society of America Bulletin* 107 (August 1995): 960–968.

27. Public Law 88–577, 88th Congress, S.4, September 3, 1964. §2(c).

28. AIRFA directs the president and federal agencies to consult with native traditional religious leaders when changing policies and procedures in order to "protect and preserve Native American religious cultural rights and practices." Public Law 95–341, §2, 92 Stat. 469 (1978) (not codified).

29. Robert Elliot, "Faking Nature," *Inquiry* 25 (March 1982): 81–93.

30. A significant body of literature exists on the restoration or remediation of damaged ecosystems. Most of the literature comes at the problem from a scientific perspective; for example, see William Jordan III, M. E. Gilpin, and J. D. Aber, *Restoration Ecology: A Synthetic Approach to Ecological Research* (Cambridge: Cambridge University Press), 1987. But there is also a body of work that highlights the philosophic dimensions of restoring nature; see Eric Katz, "The Big Lie: Human Restoration of Nature," *Research in Philosophy and Technology* 12 (1992): 231–242; and Andrew Light and Eric S. Higgs, "The Politics of Ecological Restoration," *Environmental Ethics* 18 (Fall 1996): 227–247.

31. Irene Klaver, "On Their Own Four Feet," currently under review by *Environmental Ethics*.

32. Mircea Eliade, *The Sacred and the Profane*, Willard Trask, trans. (New York: Harcourt Brace), 1968.

33. Interview with Kirk Nordstrom, USGS Water Resources Division, Boulder, Colorado, February 6, 1999.

34. Cf. Robert Weisbuch, "Six Ways to Revive the Humanities," *Chronicle of Higher Education* 45 (March 26, 1999): B4–5.

Nature and Culture

W. Scott McLean

Eldridge M. Moores

David A. Robertson

W. Scott McLean is a lecturer in two programs at the University of California, Davis (UCD)—Comparative Literature and Nature and Culture. He began his college career in fisheries biology, but changed to German languages and literature with a Ph.D. from the University of California, Santa Barbara. For the past decade he has been central to the development of UCD's Nature and Culture program. A published poet, McLean was editor of Gary Snyder's collection of interviews and talks, The Real Work *(New Directions, 1989).*

Eldridge M. Moores is Professor of Geology at the University of California, Davis. His interests include ophiolites, the development of mountains, environmental geology, and Earth history. His field experience includes all seven continents. He teaches primarily in the Geology Department but has team-taught the beginning course in the Nature and Culture program with McLean and/or David Robertson. Moores is author or co-author of more than 100 scientific publications, including two textbooks, and was the 1996 President of the Geological Society of America.

David A. Robertson is Professor of English at the University of California, Davis. He teaches in the Program in Nature and Culture and in the Graduate Group in Ecology. He is a photographer and the author of Real Matter *(University of Utah Press, 1997). His main interest in research and teaching is promoting conversations between scientists and people in the humanities.*

The Nature and Culture program at the University of California, Davis, is a bold educational experiment, striving to educate students in the essence of science, the humanities, and the arts. In this essay, the authors argue that it is only when scientific revolutions make their way into the poetry of a culture that they have truly become a part of that culture. The poetry of Gary Snyder exemplifies the centrality of geology for understanding the human spirit and its place in nature. Geology is arguably the most essential science for nonscientists to know, because of its accessibility, its role in human health and safety, its links with other sciences, and its role in the development of a sense of humility, awe, and wonder.

Geology often tells a story of beginnings. This essay is an origins narrative about a new program of great promise. The Nature and Culture (NAC) program at the University of California at Davis was established to provide a bridge between science and the humanities and arts. The program is designed to be a coherent interdisciplinary set of studies offering students the chance to explore the complex relationships existing between human cultures and the natural world. The major is a rigorous curriculum that combines courses in the natural sciences, the humanities, and social sciences, with elective courses in these and other fields.

We sat in the high ceilinged, thick-walled cavern of Kleiber Hall at the University of California, Davis. Eldridge Moores bounced down the stairs to the spare table in the front of the room, set down his briefcase and maps, and said "Good morning" to the 200 students arrayed in an amphitheater of desks above him. He went over some details of the forthcoming mid-term exam before turning to the main subject for the day, the geology of the Northern Coast Ranges of California. "It's a mess," he said. When Elohim began to create the heavens and the Earth, way back when, the Earth was "tohu wabohu." The Hebrew phrase is probably onomatopoetic and might be translated "a mess." Elohim then, as the story goes, brings order in six days by making distinctions, gathering together, setting limits, and making creatures. All this ordering is what is really unbelievable about the story in Genesis. In the beginning was a mess, and the messiness is with us still.

At present, there are three required courses in Nature and Culture itself, one lower division and two upper division. The lower-division course, entitled "Intersections of Nature and Culture," investigates nature and culture as human constructs, the importance of nature in human thought, both scientific and spiritual, and a scientific and literary view of the relation between nature and culture, including forms of observation and methods of analysis. Upper-division courses include one focusing on specific case studies and a one-week field course.

In addition to these core courses, students take a series of classes in science, literature, history, environmental studies, and a large number of electives selected from one or two thematic clusters. The latter include Human Evolution and Ecology; Human Culture and Society; Indigenous Peoples; California and the Southwest; Art and Literature; Earth and Environment; The Impact of Humans on the Environment; Environmental Law; and Policy and Planning.

A dozen instructors at UC Davis gather in the Zendo on Gary Snyder's property in the northern Sierra Nevada. It is April 1987. The meeting begins in an orderly manner. We go around the room, and one by one we say that our goal is to bridge the chasm between the humanities and the natural sciences for both our students and ourselves. Quickly, especially quickly for academics, we decided to create a new major, one that would be genuinely interdisciplinary. But it is "tohu wabohu" after lunch. That is because we have to come up with a name, and we cannot. We are a bunch of professors facile with the English language. We have among us a poet of planetary stature. For thirty minutes we try hard, then it gets wild, as we take a roller coaster ride on puns and tropes. "Okay, now," says Snyder, "this has gotten messy enough. Let's go back to one of our first suggestions, 'Nature and Culture'." "But how are we going to justify it?" some ask. "Nature and Culture has an "and" in it, and we don't believe in dualisms. Besides, people will think we mean by "nature" male and by "culture" female. Or they will think of it the other way around, with the women natural. In either case it's not good. We'll be laughed at."

"I have a solution," says Snyder. "When someone asks us why we chose a title like 'Nature and Culture' we will say, 'because we want to make problematic what our name says we are.'" With this phrase Snyder puts as well as anyone has the nature of Nature and Culture: to shake the very ground you stand on.

The three authors of this essay have been involved together in teaching the introductory course, "Intersections of Nature and Culture," for the past several years. The course typically is jointly taught by two faculty members, one each from the sciences and from the humanities. As the program is young, we are still finding our way, and the course content has varied somewhat depending on the interests and expertise of the instructors. In our collective efforts, readings have included excerpts from Descartes's *Discourse on Method*, the *Epic of Gilgamesh*, the *Bible*, the writings of Gary Snyder, John McPhee's books *Assembling California* and *Basin and Range*, as well as several of our own publications.

We strive to give students an understanding of the nature of science and its development since Descartes, the difference between experimental and historical science, and the difference between the understanding of the meaning of the word "theory" to a scientist and to the public. We endeavor also to give students an expository understanding of how changing scientific discoveries impact our understanding of who we are and our place in the world. Each of us has learned a great deal from the other two. In a real sense we are engaged in educating each other as much as educating our students.

11.1 POETRY MATTERS

It is a commonplace to speak in the history of science of a sequence of revolutions in thought: from the Aristotelian in the Middle Ages to the Copernican or "scientific" revolution of the sixteenth and seventeenth centuries, the Darwinian in the nineteenth century, and (though one may dispute whether or not Freud's work has the character of science) the Freudian revolution of the twentieth century. But if such cataclysmic upheavals in the philosophy of science mark significant turning points in epistemology and ontology, their real impact is to be found in texts that fall outside the bounds of treatises that delineate philosophic and scientific positions. *Indeed, the real reach of these revolutions is to be found in the poetries written under the aegis of profound cosmological resettings of perspective.* For example, Dante incorporates Aristotle's philosophical rigor into the *Divine Comedy*, and it is in that poem that one can truly feel the impact of Aristotle's ideas on a medieval Catholic. The poets of the late seventeenth and eighteenth centuries incorporate the advances of the new science directly into their works and irrevocably alter a generation's perspective (this is most trenchantly detailed in Marjorie Hope Nicolson's *Newton Demands the Muse*). Developments in evolutionary thought find perhaps their most heartfelt expressions in Tennyson's *In Memoriam*, a work that gives a truer gauge than any academic analysis of the ways in which the Victorians dealt with the onslaught of Darwin's evolutionary science on their spiritual moorings. And Freud's theories deeply influenced and shaped the work of countless writers in the twentieth century, beginning with Franz Kafka, forever changing our understanding of the mind's dynamics. In each of these cases, the poetic imagination engages experience in ways that mark the depth of the transformation. In each case, as Basil Wiley so incisively notes regarding Tennyson's *In Memoriam*, the poetic text, "like a piece of ritual, enacts what a credal statement merely propounds; 'this', says the poem in effect, 'is a tract of experience lived through in the light of such-and-such a thought or belief; this is what it feels like to accept it.'"

Although geology is not often given pride of place in the list of scientific revolutions referenced above, we believe that at the end of the twentieth century and the second millennium, geology indeed is the most revolutionary of all the sciences. It has, after all, undergone two revolutions in the past 50 years—those of plate tectonics and "Earth in space,"—that is, Earth's place in space, its comparison with the other planets, and the role of meteorite and/or comet impacts on processes in Earth history including evolution. These revolutions have extended the work that began with that of James Hutton, Charles Lyell, and Grove Karl Gilbert over a century ago. Today it is the geologic perspectives, concepts, and insights that continue to exercise the most profound hold on our imaginations and will, in the end, have the greatest transformative powers.

Geology does so precisely where it matters most—in the ways in which the findings of geology, as mentioned above, ultimately teach humility. In what is surely one of its most revolutionary implications, the science of unfeeling minerals and rocks leads to reverence, wonder and awe—and, in the end, to compassion. For in disclosing the forces of an inexorable world beyond our daily ken, the geologic sciences disclose the measure of the gods. That is, the geologic sciences provide a nonhuman standpoint for us to view life and experience, helping us look past the interests and perspective of the small self to the larger self that is part of nature. And when they do so, these sciences become an imaginative engagement with the world, an engagement that leads to a profound appreciation for the uniqueness and preciousness of the life of the planet—carrying us, in the end, toward an abiding compassion for all creatures bound up in time's arrow.

At a time when one often hears that the coming century will be the century of biology, one would have to add that this will be the case only if we make our own the lessons of rock's deep time; for in those lessons lies, literally as well as metaphorically, our very bedrock. One could argue just as forcefully, for the reasons outlined below, that the twenty-first century will be the century of geology.

How this is the case one can sketch by referring to the work of an American poet, our colleague Gary Snyder. For many in the sciences, most especially in the biologic sciences, Snyder is the poet of E. O. Wilson's "biophilia"—the poet of a sensibility that takes consistent reference to the genetic and species diversity of geology's multi-million-year cycles. Snyder's poems continually address our filiation with that world, seen in the image of the Buddhist "Jeweled Net" of interdependence. Snyder's is a poetry that, in the words of Joseph Baruch's poem "Prayer," is "bright with animals,/images of a gull's wing," asking "the blessing of the crayfish,/the beatitude of the birds." All of Snyder's work circles about a revisioning of that most fundamental of human markers, self-identity; and in his poetry he does nothing less than to define the ways in which our identity is to be found not in terms of our otherness, but in terms of our relations to all of nonhuman nature.

Gary Snyder is a quintessential twentieth-century American poet. Bound to the West and the landscapes of the Great Basin and Pacific Coast, his sense of culture is rooted in those vast spaces Emerson, Thoreau, and Whitman celebrated as the soul of America, an America free of the cultural bones Goethe once decried when he looked to the "New World." But Snyder's poetry is first and foremost of rock, and a full study of Snyder's two most recent books, *A Place in Space* (1995) and *Mountains and Rivers Without End* (1996), sheds crucial light on the ways this poetry is central to an understanding of geology's possible impact on our own times. Indeed, Snyder's verse enacts what the lessons of geology merely propound; this poetry offers "a tract of experience," in effect giving us poems that say what it feels like to accept the geological record.

We are well served by beginning with one of Snyder's early journal entries, this one collected in *Earth House Hold* (1969):

> The rock alive, not barren.
>   flowers lichen pinus albicaulis chipmunks
> mice even grass.

This is but one of numerous "geologic meditations" contained in Snyder's early work, the most famous of which are contained in the poems of his first volume, *Riprap*. One of the offerings in this volume, "Piute Creek," is exemplary:

> One granite ridge
> A tree, would be enough
> Or even a rock, a small creek,
> A bark shred in a pool.
> Hill beyond hill, folded and twisted
> Tough trees crammed
> In thin stone fractures
> A huge moon on it all, is too much.
> The mind wanders. A million
> Summers, night air still and the rocks
> Warm. Sky over endless mountains.
> All the junk that goes with being human
> Drops away, hard rock wavers
> Even the heavy present seems to fail
> This bubble of a heart.
> Words and books
> Like a small creek off a high ledge
> Gone in the dry air.
>
> A clear, attentive mind
> Has no meaning but that
> Which sees is truly seen.
> No one loves rock, yet we are here.
> Night chills. A flick
> In the moonlight
> Slips into Juniper shadow:
> Back there unseen
> Cold proud eyes
> Of Cougar or Coyote
> Watch me rise and go.

This poem contains a line that epitomizes Snyder's early work: "A clear, attentive mind/Has no meaning but that/Which sees is truly seen." As Snyder wrote this while working on trail crews in Yosemite setting stone pathways, this is a rock-setter's jab at Descartes's *cogito ergo sum*. In this and other poems, Snyder answers long and influential traditions in English and American poetics, for example, in Wordsworth's insistence, at the end of "1850 Prelude," that "the mind of man becomes /A thousand times more beautiful than the earth/ On which he dwells."

For Snyder, what Keats called Wordsworth's "egotistical sublime" is a fatal poetic misstep. For this grand poetic "selfhood" denies any genuine engagement beyond a focus on what we might call the "small self." Keats tried, almost singlehandedly, to turn the poetic address back toward the world itself, and in two of his most incisive letters he laid this out plainly:

> "As to the poetical Character itself," He wrote, "(that sort distinguished from the Wordsworthian or egotistical sublime;...) It is Not Itself—It Has No Self—It is every thing and Nothing—It has no character—It enjoys light and shade; it lives in gusto, be it foul or fair, high or low, rich or poor, mean or elevated. (27 October 1818)

Keats was talking about what he had earlier called "Negative Capability," when we are

> capable of being in uncertainties, Mysteries, doubts, without any irritable reaching after fact or reason. (December 1817)

A "geologic poetics" demands that we live in uncertainties, mysteries, and doubts, with a wondrous, not irritable, reaching after fact. Snyder grounds his own perspective in an imaginative reaching out to the life of rock in *Left out of the Rain* (1986):

> "Geological Meditation"
>
> Rocks suffer,
>   slowly,
> Twisting, splintering scree
> Strata and vein
>   writhe
> Boiled, chilled, form to form,
> Loosely hung over with
> Slight weight of trees,
>   quick creatures
> Flickering, soil and water,
> Alive on each other.

In this passage we see the poetic enactment of the observation recorded in an earlier journal entry—for here, in the ways in which inorganic elements take on an emotional life of their own, Snyder imagines himself into the slow suffering, the twisting, chilling life of the planet where the "quick creatures," including us, are "alive on each other."

In these poems, Snyder crafts the foundational blocks of his poetics. One of its most significant accomplishments is a perspective in which the geologic imagination, the life of rocks' million-year cycles, finds a voice. Snyder takes us deep into the landscape of the western United States and shows us how the way to the self is not so much in going deep into the self but in going deep into the world. Snyder watches carefully, attends, knows how far there is to go in learning to know the other and ultimately ourselves. Throughout his great narrative poem of the West, "Mountains and Rivers Without End," Snyder tracks the long process of attention, of listening to the rock, the wind, to write a poetry of the interstices of these contacts in what Jim Dodge has called Snyder's "poised, aggressive receptivity."

"Mountains and Rivers Without End" ends in the desert landscape and vast expanse that was once Lake Lahontan. This is, in a historically complex and abiding way, a nature poetry rooted in place and in the details of the land, with the human community at its center. And it is a geologic poetry in the sense that the landscape of rock, sand, and the million-year processes of geologic formation are active players. It is a poetry that addresses our most fundamental nature, the fact that our relationships and identities, our very own subjectivities, are found in the ephemeral substantiality of all life. Where do we find ourselves? For Gary Snyder, it is looking down the roads and across the passes and along the centuries, here in the world, to be sure—in terms of the ancient mystic traditions, finding in the immanent the transcendent. What sustains us and what really lasts are our love and our compassion, reflected in the beauty and elegance of the songs of the place, of the "foolish loving spaces/full of heart," where "All art and song/is sacred to the real."

11.2 GEOLOGY MATTERS

Although geology is a mature and complex science with roots in both natural and experimental fields of study, it is an ideal subject for nonscience students to learn because of four intrinsic factors: Geology gives a sense of place: it offers a historical perspective; it makes other sciences, such as physics, chemistry, and biology, more accessible to the nonscientist; and it offers necessary information for human welfare and safety. We briefly discuss each of these factors in turn.

A SENSE OF PLACE

While Gary Snyder's poetry may provide appropriate geologic grounding to many of us today, humans have had an intrinsic need, throughout their entire history, to be connected to their surroundings and to understand the world around them. Much of this need involves an understanding of the land and where it comes from, which intrinsically involves considerations of geology. It is perhaps not surprising how strongly descriptions of the land and its origins resonate with nongeologists, because human writings from the earliest myths and narratives have focused on the land, and humans' relationship to it (e.g., Torrance 1998). The land and where we live shape our perceptions and our philosophy.

Humans have historically possessed a fundamental need to feel that the land is solid and firm beneath their feet. Throughout most of human history, any infrequent geologic event that upsets that feeling, such as a volcanic eruption or an earthquake, has been viewed as the ultimate wrath of a vengeful god or gods. With the advent of the plate tectonic revolution in the late twentieth century, we now know, of course, that in the long run the entire Earth is unstable, but individual events tend to occur on a time scale that is long with respect to an average human lifetime. These events are the principal forces that shape the landscape. Or, as Snyder puts it in *Mountains and Rivers Without End"* (p. 145)

> ten million years ago an ocean floor
> glides like a snake beneath the continent crunching up
> old seabed till it's high as alps.

A SENSE OF PERSPECTIVE

Two factors of paramount importance for the average person to know are the sense of the length of geologic time, and the slow process of geologic events through this time. Analogies and metaphors are especially important in giving the nonscientist a sense of geologic time or of astronomical distance, which in turn give meaning to life. All of us have our favorite metaphors. We have found it useful to equate years to millimeters or to seconds. In millimeters, all of recorded history is contained within 10 meters, and the age of Earth spans the North American continent from the Atlantic to the Pacific; in seconds, all of recorded history is about three hours, and Earth's age is about 150 years!

In such discussions, it is important also to give a sense of the pace of geologic events, such as major earthquakes and their recurrence intervals, the frequency of "hundred-year" floods, and the artificiality of the latter concept. The usual outcome of this sort of discussion is the development in students of a sense of humility, similar to that invoked by the poetry of Gary Snyder, and a renewed sense of the meaning of the statement "the Earth was not made for us, we were made for the Earth."

LINKS WITH OTHER SCIENCES

Geology is holistic and cross-disciplinary. It uses the principles of the "pcb's"—physics, chemistry and biology—to understand the workings and history of Earth and its life and, in the future, that of other planets as well. Because these fields tend to be reductionist, specialized, and hierarchical, they include many "little islands of near conformity surrounded by interdisciplinary oceans of ignorance" (Ziman 1996), and they are difficult for the nonscientist to approach. Because geology utilizes all three fields, it gives the nonscientist an understanding of how one science relates to another, and the "relevance" of the various aspects of scientific knowledge. Because geology is involved in the generation of a narrative, it makes the other sciences more accessible and "demystifies" or "demythologizes" them. In doing so it makes them in turn more understandable and approachable and less fearful to the average person.

HUMAN WELFARE AND SAFETY

We live on an increasingly crowded planet, one with finite resources. In the twenty-first century the continued increase in population, the revolution in communication, and development of a "global community," the aspirations of everyone with access to a TV set to develop a "Western" or "American" consumption-based lifestyle, coupled with the finiteness of Earth's resources, especially water and energy, mean collectively that we are going to be approaching the ultimate limits of Earth's carrying capacity (e.g., Youngquist 1997; Moores 1997). Humanity will need to make choices, either to descend into war or to work together. Several recent estimates predict that the world petroleum production will peak sometime in the next 50 years (e.g., Kerr 1998; Youngquist 1997). Water is already scarce for about one-third of the world's people, and will become increasingly so in the twenty-first century. Growing populations and the attendant need for added food production and area for housing mean collectively that there will be increasing pressure on land. Urbanization will continue to expand into its geologically unstable or otherwise hazardous regions.

To minimize this increasing threat to human safety, it becomes important to factor geologic information into policy discussions and decision making. This will not be easy, because it will require a wrenching change in the mind-set on the part of politicians and policy professionals, most of whom have little or no contact with the geological sciences.

We need individuals who are educated to bridge the gap between the sciences and the humanities. We find that the Nature and Culture (NAC) program does precisely this. Moreover, NAC provides a newly configured liberal arts education that embodies the very best of what has always been the genuine interdisciplinary nature of truly creative minds. In a very real sense it teaches the student how to stand back and think creatively. We are also finding that our students thus educated are attractive to a diverse range of employers; NAC graduates have gone into such fields as population control, fisheries, law, and investment banking.

We need to make a long-term, sustained effort to cultivate "Earth awareness." We collectively need to begin to provide scenarios to decision-makers, rather than attempt predictions. No matter what their ultimate professional position, Nature and Culture graduates will have an understanding of Earth and its long-term processes that we will need to meet the challenges of the future.

11.3 FINAL THOUGHTS

It was a tolerably warm September day. Students in NAC 180, the senior fieldwork capstone course, were driving west out of Davis heading toward the upper watershed of Putah Creek. It was a day-long trip with Moores. Five miles out of town we stopped in front of a gentle, low hill—well, hardly a hill, more like a mound. He said, "This rise in the ground is the easternmost structure of a series of folds over hidden or 'blind' thrust faults along the western side of the Great Valley. You can think of it as an incipient mountain in the Coast Range." We pulled over again before the town of Winters, hard hit by an earthquake in 1892. "We are standing on top of 40,000 feet of fill, or on top of 20,000 feet doubled by imbricate structure. It is hard to tell." At the next stop along the juncture of the Tehema Formation's debris derived from the Coast Range uplift and the uplifted Cretaceous of the eastern Great Valley Sequence, a small valley is strewn with boulders of basalt. "They have no known source. I think it is possible they are the tail of the Columbia Plateau. Now how would you go about testing a harebrained idea like that?"

We gathered in Stebbins Cold Canyon, one of the University of California's Natural Reserves, to discuss the area with a geology student writing an honors thesis. His project was to do a geological map of the Reserve. Moores said to him, "What you want to do is pull out the lithographic units and see if they are real: if they are real." Later in the trip he commented to all of us, as he looked out over the terrain upcanyon, "This is upper Cretaceous, unless it isn't."

It is the "messiness" made by joining

> *"I think...."*
> *"It is hard to tell...."*
> *"How would you go about testing...?"*
> *"If they are real...."*
> *"Unless it isn't...."*

that is at the heart of the Nature and Culture program. It is the distance from one's own views, the self-reflection, the questioning, the uncertainty, the throwing it all back on the students, all these together, that make NAC what it is. The following lines from Gary Snyder's poem "The Mountain Spirit" (published in *Mountains and Rivers Without End*) capture the essence of the humility one feels when contemplating the whole Earth:

> *Walking on walking,*
> *under foot earth turns*
>
> *Streams and mountains never stay the same*
> *Walking on walking*
> *under foot earth turns*
>
> *Streams and mountains never stay the same.*

REFERENCES

Kerr, R. A., 1998, "The next oil crisis looms large—and perhaps close." *Science*, v. 281, pp. 1128–1131.

Kovacs, M. G., Transl., 1989, *The Epic of Gilgamesh*. Stanford, CA: Stanford University Press.

McPhee, J., 1993, *Assembling California*. New York: Farrar, Straus, and Giroux.

McPhee, J., 1981, *Basin and Range*. New York: Farrar, Straus & Giroux.

Moores, E. M., 1997, "Geology and culture: A call for action," *GSA Today*, v. 7, no. 1, pp. 7–11.

Snyder, G., 1986, *Left out in the Rain*. San Francisco, CA: North Point Press..

Snyder, G., 1996, *Mountains and Rivers Without End*. Washington, DC: Counterpoint.

Snyder, G., 1995, *A Place in Space: Ethics, Aesthetics, and Watersheds*. Washington, DC: Counterpoint.

Snyder, G, 1969, *Earth House Hold: Technical Notes & Queries to Fellow Dharma Revolutionaries*. New York: New Directions.

Sutcliffe, F. E., Transl., 1968, *Discourse on Method and the Meditations*, by René Descartes. London: Penguin.

Torrance, R. M., 1998, *Encompassing Nature: A Sourcebook*. Washington, DC: Counterpoint.

Youngquist, W., 1997, *GeoDestinies The Inevitable Control of Earth Resources Over Nations and Individuals*. Portland, OR: National Book Co.

Ziman, J., 1996, "Is science losing its objectivity?" *Nature*, v. 382, pp. 751–754.

PART III

PHILOSOPHIC APPROACHES TO THE EARTH

1 2

Earth Religions, Earth Sciences, Earth Philosophies

Carl Mitcham

A member of the faculty at the Colorado School of Mines, Carl Mitcham's schol-arly research has been in the philosophy of science and technology as well as in the interdisciplinary field of science, technology, and society studies. His fundamental concern has been the attempt to understand, from what he calls a historico-philo-sophical perspective, the structures of the contemporary technoscientific lifeworld. Among his most representative publications are Philosophy and Technology: Readings in the Philosophical Problems of Technology, *co-edited with Robert Mackey (Free Press, 1972; paperback 1983), and* Thinking Through Technology: The Path Between Engineering and Philosophy *(University of Chicago Press, 1994). His current work focuses on the ethics of science and technology.*

Mitcham's contribution to this collection begins by situating his reflection within a historico-philosophical framework that, although it has been subjected to extensive criticism, remains broadly popular: that of the differences among reli-gion, philosophy, and science, in which science is taken as the highest form of knowing. Mitcham centers attention on the place of the geosciences within this schematism. His appeal for recognizing the role of the Earth in the history of reli-gions, and for a more metaphysically sensitive appreciation of the history of geol-ogy, sugggests a critical assessment of contemporary environmental philosophies. In a challenge to contemporary environmental philosophy, Mitcham concludes with the citation of "Declaration on Soil" from the radical social critic Ivan Illich, who is most well known for his withering criticisms of the counterproductivity of such institutions as schools and the healthcare system–e.g., Deschooling Society–*(Harper & Row, 1971) and* Medical Nemesis *(Pantheon, 1976). But since the 1980s Illich's orientation has been more toward the practice of a historical archeology that makes common cause with Mitcham's own approach.*

Is it possible to say anything useful about the relation among religions, philosophies, and sciences of the Earth, other than that any science of the Earth—whether earth be taken as rock, soil, or a global system—would constitute the only true and valid geological knowledge? To this mildly rhetorical question the prejudgment, that is, the prejudice, of our time is a quick negative response. Neither religion nor philosophy, but only science, is commonly thought able to present us with relevant and reliable knowl-edge about the Earth. Indeed, according to a positivist schematism promoted by the nineteenth-century French philosopher August Comte, who actually argued this posi-tion, the history of knowledge may be described as one in which religion is followed by philosophy is followed by science. Philosophy grows out of and replaces religion; science grows out of and replaces philosophy.

In partial confirmation of this modernist interpretation of history and knowledge, Earth religions have indeed preceded the Earth sciences. In partial refutation of Comte, however, Earth philosophy does not yet or only just barely exists. Any simple substantiation of these two claims is complicated by the problematic character of all three basic terms, but may nevertheless be explored as an exercise in historico-philosophical orientation.

The assumption of this exercise is that the effort to understand relations among past, present, and future—not to mention among religion, philosophy, and science—is more constitutive of human experience than the simple positivist belief that science is the culmination of intellectual history. Such an exercise thus requires some minimal willing suspension of commitment to the idea of a positive, progressive chronology, in favor of a dialogue among alternative ways of being in the world. The hypothesis of this essay, in its questioning of Comte's typically modern schematism, is precisely that history may be read not as a triumphantalist march to the present but as a long and often divergent encounter among different understandings about the nature and meaning of the world and the place of human beings within it.

12.1 RELIGIONS OF THE EARTH

The word "religion" comes from the Latin *religare*, meaning to tie (a boat or ship to the dock), to fasten behind (a person's hair), or more generally to be connected with. That with which religion connects is the sacred, which is in turn characterized by being separated from the everyday, as transcending what is at once both common and necessary. Yet at the same time, and paradoxically, what is today thought of as religion was for most of human history not seen as something separate from, say, politics, economics, or art. Instead, religion pervaded human experience, especially human experience of the Earth.[1]

Indeed, rituals and hierophanies or revelations of the Earth are among the most primordial forms of the religious life and experience. In the words of a Homeric hymn, "It is the Earth I sing, securely enthroned, the mother of all things, venerable ancestress feeding upon her soil all that exists" ("To Earth," line 1). According to Hesiod, it is from Earth that there sprang not only mortals but also the gods. "Earth herself first gave birth to a being equal to herself, who could overspread her completely, the starry heaven" (Hesiod, *Theogony*, V, 126). From the union of Earth and Sky came the original families of the gods, who married and gave birth to other gods, and whom human beings imitate in their own earthly marriages. According to a model provided by the *Upanishads*, human marriage partakes of hierogamy as the husband declares to his wife, "I am Heaven, thou art Earth." The marriage of Dido and Aeneas takes place in the middle of a violent storm, their union coinciding with a riotous mixing of the elements (Virgil, *Aeneid*, VI, 160).[2]

Even in the Jewish, Christian, and Islamic traditions, in which the creator-God transcends the world, Earth is created before humans. Indeed, according to the Hebrew Scriptures, human beings were formed from the soil of the Earth. Their partaking of the fruit of the tree of knowledge, that which springs from and may thus have been thought to lift them above Earth, paradoxically returns human beings more firmly to it. "Soil thou art, and unto soil thou shall return" (Genesis 3:19).

It is the earth as soil which is also invoked and cultivated in agricultural rites, and which is seen as a source of regeneration through descent into caves and burials in the soil of a homeland. In the words of a Zuñi prayer chant to the Earth divinity,

> May the rain-makers water the Earth Mother that she may be made beautiful to look upon. May the rain-makers water the Earth Mother that she may become fruit-ful and give to her children and to all the world the fruits of her being, that they may have food in abundance. May the Sun Father embrace our Earth Mother that she may become fruitful, that food may be bountiful, and that our children may live the span of life, not die, but sleep to awake with their gods.[3]

Earth's life-giving power is manifest not only through agriculture. Orpheus, Jesus, and Dante all descend into Earth's rocky depths in a process of death and regeneration.

Beneath that animal and vegetable effulgence covering Earth's surface, and coordinate with gnostic descent into what the Romans knew as *Terra Mater* (Earth Mother), is the seminal life of stones. When Zeus poured water on the Earth and flooded it, Deucalion and his wife, Pyrrha, escaped in an ark constructed on the advice of his father, Prometheus. After floating for nine days the ark came to rest on Mt. Parnassus. Deucalion came out of the ark and offered a sacrifice to Zeus, who granted him a single wish. Deucalion chose to wish for other humans, and was directed to throw stones over his head. When he did so they became the seeds of men; stones thrown over her head by Pyrrha became women, and thus they repopulated the Earth.

The bones of the Earth Mother constitute archetypical images of that which is real, are seeds of human beings, and mark out in mountains, stella, and stone tablets multiple presences of the gods and the solidities of human life. Long before St. Augustine's restlessness found rest in God, or Immanuel Kant appealed to the twin stabilities of the starry heavens above and the moral law within, archaic peoples found firmness in the sacred Earth, its rocks and its soils. They understood themselves as related to a terrestrial place, and defined themselves in terms of totems taken from nature. Here is a clan of the bear or the wolf; there is a tribe of the forest or a people of the desert.

The Earth and its stones are also and at once both sacred and sexual. Streams, caves, and the galleries of mines are described as the vagina of the Earth Mother; everything in the Earth's belly is in a state of gestation. Ores extracted from mines, for instance, are embryos:

> they grow slowly as though in obedience to some temporal rhythm other than that of vegetable and animal organisms. They nevertheless do grow—they "grow ripe" in their telluric darkness. Their extraction from the bowels of the earth is thus an oper-ation executed before its due time. If they had been permitted the time to develop (i.e., the *geological rhythm of time*), the ores would have become ripe metals, having reached a state of "perfection."[4]

The Earth is shot through with life; there is no strictly or completely dead matter. It is not that the distinction between the organic and the inorganic has not yet been made; such a distinction is rejected as not true to the experience of a cosmos animated by seasons and the most subtle of changes, not unlike those we experience in our own lives. The ur-geologist or miner is thus also and necessarily a priest who must be set apart from society, who is both contaminated and purified through his interventions, and must offer sacrifices as part of his intercourse with a vibrant Earth.

It is an alienation from this Earth as sacred Mother that clears the way for a technological conquest and scientific examination of the Earth. Carolyn Merchant summarizes her analysis of this transformation in *The Death of Nature*: "Between the sixteenth and seventeenth centuries the image of an organic cosmos with a living female Earth at its center gave way to a mechanistic world view in which nature was reconstructed as dead and passive, to be dominated and controlled by humans."[5] That is, the Scientific Revolution, which treats nature as a machine to be known by means of experimental manipulation, yields knowledge that in its very essence gives to human beings the power to control and manage the world.[6] Crucially, modern natural science and the coordinate manifestation of an Industrial Revolution rest on a fundamental and thoroughgoing rejection of any religion of the Earth that would identify any aspect of nature as a sacred reality limiting human action.

12.2 SCIENCES OF THE EARTH

Out of the religions of the Earth there grew no philosophies of the Earth. Geology is conspicuous by its absence in, for instance, Plato, Aristotle, Cicero, Augustine, and Thomas Aquinas. The most that can be found are isolated mentions of geological phenomena such as volcanic explosions and earthquakes, philosophically interpreted myths such as in Plato's *Timaeus*, and a few short, isolated treatises such as Theophrastus's *De lapidibus* ("On stones"), which merely collects otherwise disparate information about rocks, minerals, and fossils. Premodern natural histories such as those of Herodotus, Strabo, or Pliny the Elder collect and narrate stories of diverse natural phenomena, with only occasional and modest speculative interpretations of their nondivine causes.

Philosophy begins, according to Aristotle, when discourse about the gods is replaced by discourse about the inner natures as the causes or reasons of things.[7] Socrates, according to Cicero, took such discourse, which focused in the first instance on the heavens, and brought it into the cities to deal with human affairs.[8] Although the typically modern assumption is that classical philosophers failed to develop a science of the Earth because the Earth was something to be looked down upon, was merely matter, it is perhaps more true that the phenomena of the Earth were excluded from the pursuit of the systematic knowledge of nature precisely so that they might be left in the care of those who discourse on the gods. Some things continued to be acknowledged, even by ancient philosophers, as appropriate to the gods.

The exclusion of the gods that Merchant argues is a precondition for the rise of modern natural science and technology in general—that is, in the first instance, for physics and chemistry—is also and prominently associated with the rise of geology, in the second wave of modern scientific disciplines that arose in the eighteenth and nineteenth centuries. The opening pages of what is generally recognized as the first treatise in scientific geology, for instance, constitute a decisive break with all religions of the

Earth. According to the very first sentence of James Hutton's "A Theory of the Earth" (1788), "When we trace the parts of which this terrestrial system is composed, and when we view the general connection of these several parts, the whole presents a machine of a peculiar construction by which it is adapted to a certain end."[9] This image of the earth as a machine is repeated like a mantra in the opening pages of Hutton's text:

> We have thus surveyed the machine in general, with those moving powers, by which its operations, diversified almost *ad infinitum*, are performed. Let us now confine our view, more particularly, to that part of the machine on which we dwell, that so we may consider the natural consequences of those operations which, being within our view, we are better qualified to examine. This subject is important to the human race, to the possessor of this world, to the intelligent being Man, who foresees events to come, and who, in contemplating his future interest, is led to enquire concerning causes, in order that he may judge of events which otherwise he could not know.[10]

For Hutton, the earth is neither a divinity nor is it to be understood by reference to the divine. It is a machine, and as such may be manipulated at will. His comparisons echo the mechanistic interpretation of the early modern philosopher Thomas Hobbes (1588–1679), who had previously bridged the distinction between life and machine by seeing, on the one hand, machines as life-like and, on the other, human beings as machines. In Hobbes's words,

> Seeing that life is but a motion of limbs, the beginning whereof is in some principal part within, why may we not say that all *automata* (engines that move themselves by springs and wheels as does a watch) have an artificial life? For what is the *heart* but a *spring*, and the *nerves* but so many *strings*, and the *joints* but so many *wheels* giving motion to the whole body....[11]

Moreover, in Hutton's deft reference to the "human race" as "the possessor of this world," he further echoes and assumes the commitments of Francis Bacon (1561–1626) to the endowment of human life "with new discoveries and powers" and of René Descartes (1596–1650) to make human beings "masters and possessors of nature."[12] With Hutton, the sixteenth and seventeenth centuries have been reborn and applied to the earth on which we stand.

Mirroring his profession as a physician—but as a physician of Hobbes' human machine—Hutton thus argues that the new knowledge of the Earth is not simply a theory. It is a practical theory, even a medicinal theory. "Is this world to be considered thus merely as a machine, to last no longer than its parts retain their perfect position, their proper forms and qualities?" he asks, rhetorically.[13] The answer is obviously not.

> In what follows, therefore, we are to examine the construction of the present earth, in order to understand the natural operations of time past; to acquire principles, by which we may conclude with regard to the future course of things, or judge of those operations, by which a world so...ordered, goes into decay; and to learn, by what means such a decayed world may be renovated, or the waste of habitable land upon the globe be repaired.[14]

No longer is the order of the Earth thought to stand on its own, like a god, or left in the hands of God. Instead, it is placed in, even calls for, the hand of humanity—adumbrating contemporary aspirations to Earth systems' management.

The second great founder of geology, Charles Lyell, reiterates Hutton in regard to both these commitments to de-divinization and human service. According to Lyell's *Principles of Geology* (1830),

> Geology is the science which investigates the successive changes that have taken place in the organic and inorganic kingdoms of nature; it enquires into the causes of these changes, and the influence which they have exerted in modifying the surface and external structure of our planet. By these researches into the state of the earth and its inhabitants at former periods, we acquire a more perfect knowledge of its *present* condition, and more comprehensive views concerning the *laws* now governing its animate and inanimate productions.[15]

Geology is not a cosmogony or story of origins but a history of causal relations that may increase human productivity and power[16]—indeed, on some interpretations, geology may even be said to call forth such productivity and power.

The progressive development of the sciences of the earth since the late eighteenth and early nineteenth centuries has, as Lyell himself predicted, been nothing short of staggering.[17] Under the weight of this development, the Earth sciences have been divided into the geologic sciences (including mineralogy, physical geology, petrology, structural geology, geomorphology, geophysics and geochemistry, and more), the hydrologic sciences (hydrology, oceanography, limnology, glaciology), and the atmospheric sciences (meteorology, climatology, aeronomy). The history of this development has been variously described, at different stages, as an argument between Neptunists (who maintained that an original global ocean was the primary influence on geological formation) and Plutonists (who viewed heat as the primary force driving geological change), as the evolution of a positivist science, as a transition from mineralogy to geology, and as the triumph of key concepts such as uniformitarianism, deep time, or plate tectonics.[18] The most recent history of the Earth sciences, with the development of geoinformation systems and computer modeling, might also be described as using meta-technology to create a meta-Earth.[19]

Only the last of these narratives begins to come to grips with the fundamental paradox or philosophical challenge posed by modern geology. The Earth sciences developed in order to increase human power and control, a goal that assumed the special place of human beings in a natural order. From the perspective of geological forces and time, however, as these have been revealed by the Earth sciences, one may question whether human beings deserve any special power and control. Yet given their insignificance, why shouldn't human beings do just as they please?

Geology places the earth in human hands, and for more than a century served as handmaid of the human exploitation of its resources. More recently the geosciences have attempted to inventory limits to certain resources (such as oil or gas). Does this imply an end to exploitation, or simply call for an ever more thoroughgoing exploitation in the form of planetary management? To those who, in our democracy of desire, argue that limits should be met with political restraint and the preservation of ecologies, others patiently explain that change and instability are fundamental to the geological system, so that the only valid arbiter of human action is a sufficiently far-reaching and participatory economy of self-interest. Turbo-capitalist global entrepreneurs find in resource limits as many opportunities as do environmentalists.[20]

12.3 PHILOSOPHIES OF THE EARTH

There have been no philosophies of geology, much less of the earth. The first article in a major philosophical reference work that even used the term did not appear until the end of the twentieth century. In her article, "Philosophy of Geology," Rachel Lauden nevertheless accurately noted that philosophical reflection on geology almost exclusively consisted in using the history of the rise of geology to analyze scientific change.[21] David Kitt's attempt to apply logical empiricist philosophical principles to geology—that is, to read geology as, like physics, formulating logically well-formed laws that cover and therefore explain a multitude of empirical data—is virtually the lone exception. Yet Kitt's attempt was already dated by the time his book—*The Structure of Geology*—was published, given that Thomas Kuhn's *The Structure of Scientific Revolutions*, which points up the historical inadequacies of logical empiricism, appeared on the scene 15 years earlier.[22] Bypassing Comte's sequence, geological science has clearly emerged without the prior emergence of any full-bodied philosophy of the Earth—that is, any attempt to articulate an integrated epistemology, metaphysics, and ethics of the Earth.

It might plausibly be argued, however, that environmental ethics constitutes a kind of philosophy of the Earth sciences—especially insofar as environmental ethics has evolved into environmental philosophy. Indeed, the very replacement of the term "geology" with "Earth sciences"[23] took place under the aegis of the rise of the contemporary environmental movement, galvanized in large measure by the 1962 publication of conservation biologist Rachel Carson's *Silent Spring* and photographs by Apollo astronauts of the planet from space.[24] (This same "global" sensibility has at the same time been lacking in appreciation of previous environmental initiatives in North American history, not to mention those in the histories of countries and cultures further afield.)[25] But if analytic philosophies of geology such as Kitt's limits themselves to narrow epistemological issues, environmental ethics tends to spend itself in broad but largely impotent moralism.

Some approaches to environmental ethics have, it is true, undertaken an analytic philosophical critique of ecology or ecosystem biology, arguably another branch of the Earth sciences.[26] The more common path, however, has been either to treat the term "Earth sciences" as a neutral but rhetorically useful term or to take the rhetoric seriously and to adopt, usually more implicitly than explicitly, a romantic reading of geology as a kind of caring for the Earth.

The romantic reading may itself be read as an attempt to reaffirm or recover, in the midst of the Earth sciences, some dimension of archaic earth religions. These religions were concerned not with theoretically unified, empirically controlled, and power-centered knowledge, but with what may be incompletely characterized as affectively ethical knowledge. That is, on the ground that human beings were of the Earth, archaic peoples struggled to call attention to and to articulate dimensions of this belonging, to invite themselves to practice and to think their belonging, to limit those ways of life that would interfere with and infringe upon such belonging, and to intensify the belonging in appropriately human ways—that is, through narrative and poetry.

For the geological sciences as sciences there is an immediate stepping outside of this spiritual–ethical framework. On the assumption that human beings are not of the Earth, that they do not in fact fit in with earthly ways, that as a species humans thus have a right to seize power over even the Earth on which they live, a power that they nevertheless lack, modern geology carried forward that historical break that constitutes modernity. From Nicolo Machiavelli's politics of power and Isaac Newton's celestial mechanics to Hutton's terrestrial medicine, progressive waves of modernity have intensified human power and extended human control—ever exploring, exploiting, and managing the Earth.

This is not to deny that there have been many pious geologists. Indeed, Hutton's follower, Lyell, makes a case for geology as a positive influence on religion, decontaminating it of mystical and superstitious elements.[27] But the piety of the modern geological sciences remains an individual and private add-on, not an inner and publicly acknowledged constituent.

The contemporary environmental movement, insofar as it seeks to transcend strictly medicinal commitments—that is, commitments to make the Earth more useful and satisfactory to humans, leaving the desires of humans to play themselves out as they will in a global economy of scarcity and entertainment—may thus constitute an attempt to recover premodern bonds between humans and the Earth. But such bonds are only weakly articulated in intellectual calls for deep ecology or nonanthropocentric environmental ethics. It remains unclear whether such philosophical discourse about the Earth is either appropriate or possible. In multiple senses, for philosophies of the Earth, the Earth does not matter.

12.4 THE EARTH DOES NOT MATTER

At the core of the absence of a philosophy of the Earth, paradoxically, there may be too much talk about the Earth. In all this talk, the Earth simply does not matter—does not reveal itself as such. The Earth does not matter because its ability to matter has, from Hutton and Lyell on, been methodologically restricted to mere matter. The Earth does not matter because, in the Earth sciences, it has been transformed into a problem for multi- and interdisciplinary attack. The Earth does not matter because under the gaze of Apollo astronauts, integrated geoinformation systems, and computer modeling it has ceased to exist as the ground on which we stand. The Earth does not matter because from agribusiness investment and biotechnical design to climatological management, the Earth has been transformed into a meta-Earth. In such circumstances it may be precisely that the very rhetoric of the Earth is itself an impediment to our attempt to think the Earth.

Consider, for instance, the words of the Amerindian prophet Smohalla, who refused to till the soil:

> You ask me to plow in the earth! Shall I take a knife and tear my mother's bosom? Then when I die she will not take me into her bosom to rest.
> You ask me to dig for stone! Shall I dig under her skin for her bones? Then when I did I can not enter her body to be born again.
> You ask me to cut grass and make hay and sell it, and be rich like the white men! But how dare I cut off my mother's hair?...
> We simply take the gifts that are freely offered. We no more harm the earth than would an infant's fingers harm its mother's breast. But the white man tears up large tracts of land, runs deep ditches, cuts down forests, and changes the whole face of the earth.[28]

How is it possible to speak and to think in this manner? What Smohalla speaks mythically and poetically about the earth as specifically placed soil—unifying that which philosophy calls epistemology (a theory of knowing), metaphysics (insight into reality), and ethics (understanding of the good)—is this possible to think today? More than a century later is there anyone who at least echoes such an archaic stance?

In contrast to the way the story is commonly told, the contemporary environmental movement has many predecessors. Among these predecessors is the organic gardening movement, founded in the 1930s by J.I. Rodale and continued by his son, Robert Rodale. In December 1990, following the fall of Communism, Robert Rodale was in Moscow arranging to publish a Russian-language edition of his magazine, *Organic Gardening*. On his way back to the United States, Rodale planned to stop off in Germany to participate in a philosophical discussion on organic agriculture with his friend Ivan Illich and others. However, Rodale was killed in an automobile accident on his way to the airport. Stunned by his sudden death, the discussion participants drew up—in the spirit of Rodale—the following document, a "Declaration on Soil":

> The ecological discourse about planet earth, global hunger, threats to life, urges us to look down at the soil, humbly, as philosophers. We stand on soil, not on earth. From soil we come, and to the soil we bequeath our excrements and remains. And soil—its cultivation and our bondage to it—is remarkably absent from those things clarified by philosophy in our Western tradition.
>
> As philosophers, we search below our feet because our generation has lost its grounding in both soil and virtue. By virtue, we mean that shape, order, and direction of action informed by tradition, bounded by place, and qualified by choices made within the habitual reach of the actor; we mean practice mutually recognized as being good within a shared local culture which enhances the memories of a place. We note that such virtue is traditionally found in labor, craft, dwelling, and suffering supported, not by an abstract earth, environment, or energy system, but by the particular soil these very actions have enriched with their traces. And yet, in spite of this ultimate bond between soil and being, soil and the good, philosophy has not brought forth the concepts which would allow us to relate virtue to common soil, something vastly different from managing behavior on a shared planet.
>
> We were torn from the bonds to soil—the connections which limited action, making practical virtue possible—when modernization insulated us from plain dirt, from toil, flesh, soil, and grave. The economy into which we have been absorbed—some, willy-nilly, some at great cost—transforms people into interchangeable morsels of population, ruled by the laws of scarcity.
>
> Commons and homes are barely imaginable to persons hooked on public utilities and garaged in furnished cubicles. Bread is a mere foodstuff, if not calories or roughage. To speak of friendship, religion, and joint suffering as a style of conviviality—after the soil has been poisoned and cemented over—appears like academic dreaming to people randomly scattered in vehicles, offices, prisons, and hotels.
>
> As philosophers, we emphasize the duty to speak about soil. For Plato, Aristotle, and Galen it could be taken for granted; not so today. Soil on which culture can grow and corn be cultivated is lost from view when it is defined as a complex subsystem, sector, resource, problem, or "farm"—as agricultural science tends to do.
>
> As philosophers, we offer resistance to those ecological experts who preach respect for science, but foster neglect for historical tradition, local flair and the earthy virtue, self-limitation.

Sadly, but without nostalgia, we acknowledge the pastness of the past. With diffidence, then, we attempt to share what we see: some results of the earth's having lost its soil. And we are irked by the neglect for soil in the discourse carried on among boardroom ecologists. But we are also critical of many among well-meaning romantics, Luddites and mystics who exalt soil, making it the matrix, not of virtue, but of life. Therefore, we issue a call for a philosophy of soil: a clear, disciplined analysis of that experience and memory of soil without which neither virtue nor some new kind of subsistence can be.[29]

Ivan Illich's plea for "a clear, disciplined analysis of that experience and memory of soil" thus constitutes a kind of summons, once again, to develop a philosophy of the Earth. In contrast to other such summonses, however, it directs attention not at something like the Earth as a system of image and economy, but the ground on which we walk. Like Socrates, Illich would bring philosophy down from the heavens to dwell among humans and those convivial habits that run the gamut from work and friendship to feast and suffering.

12.5 EPILOGUE

Can we say anything more about the relation among religions, philosophies, and sciences of the Earth than that the sciences of the Earth constitute the only true and valid geological knowledge? From the point of view of archaic religions of the Earth the question can almost be reversed: What can the sciences of the Earth tell us that we do not already know? Given both perspectives, can a philosophy of the Earth do more than simply side with one position or the other? Is the charge of philosophy only to defend the present, or to attack it? Why is there no philosophy of the Earth that would mediate between these two positions?

The rise of modern natural science is one of the greatest events in human history—and one of the most mysterious. Just as we know only partially the causes and complexities behind the rise of the early great urban civilizations of Egypt, Mesopotamia, India, and China, so we only begin to appreciate the causes of the rise of modern natural science and its full implications. In both cases, what knowledge we have must not be allowed to obscure what we do not know. A too quick defense of science or of the great urbanizations of human history covers over the richness of our own past—a richness that may properly be called upon to enlarge our future.

Geology, as part of natural science, which is itself allied with the new urbanization of modernity, partakes in the greatness of these historical events—and in their mysteries. To become knowledgeable, it is not possible simply to turn away from such events, to direct one's attention elsewhere. To fail to pursue appreciation of the powers of science and of its losses is to fail to exercise our humanity in our own historical period. Everywhere we turn with open eyes in our technoscientific world we are reminded again and again of what science has given and of what science has taken away. Rather than being hyperactive cheerleaders for the future or sullen critics of the present we must attempt to think it, to recognize ever more clearly what has happened and what is at stake—in part, through a studied dialogue with ancestors and alternatives, which it has been the modest aim of this essay to stimulate.

NOTES

1. For classic introductions to the study of religion, see, e.g., William James, *The Varieties of Religious Experience* (New York: 1905); Emile Durkheim, *The Elementary Forms of Religious Life: A Study in Religious Sociology*, trans. J.W. Swain (New York: Macmillan, 1915); Bronislaw Malinowski, *Magic, Science and Religion, and Other Essays* (Boston: Beacon Press, 1948); Mircea Eliade, *Patterns in Comparative Religion*, trans. Rosemary Sheed (New York: Sheed and Ward, 1958); and Mircea Eliade, ed., *Encyclopedia of Religion*, 16 vols. (New York: Macmillan, 1987). The entries on "Earth, Earth-Gods" in James Hastings, ed., *Encyclopedia of Religion and Ethics* (New York: Scribner's, 1908), vol. 5, pp. 127–131; and on "Earth" in the *Eliade Encyclopedia of Religion*, vol. 4, pp. 534–541, provide basic references for the discussion that follows.

2. Mircea Eliade, *Cosmos and History: The Myth of the Eternal Return* (New York: Harper & Row, 1959), pp. 22–23.

3. Matilda Coxe Stevenson, "Ethnobotany of the Zuni Indians," *Annual Report of the Bureau of American Ethnology*, vol. 30:1908–1909 (Washington, DC: Government Printing Office, 1915), p. 37.

4. Mircea Eliade, *The Forge and the Crucible*, trans. Stephen Corrin (New York: Harper & Row, 1962), p. 42.

5. Carolyn Merchant, *The Death of Nature: Women, Ecology, and the Scientific Revolution*, 2nd ed. (New York: Harper & Row, 1990), p. xvi.

6. For an extended argument to this effect, see Hans Jonas, "The Practical Uses of Theory," in Jonas, *The Phenomenon of Life: Toward a Philosophical Biology* (New York: Harper & Row, 1966), pp. 188–210.

7. See, e.g., Aristotle, *Metaphysics* I, 7 (988b27); III, 4 (1000a9); XII, 6 (1071b27); XII, 10 (1075b26); and XIV, 4 (1091a34).

8. Cicero, *Tusculan Disputations* V, iv, 10–11.

9. James Hutton, "A Theory of the Earth; or an Investigation of the Laws Observable of the Composition, Dissolution, and Restoration of Land Upon the Globe." *Transactions of the Royal Society of Edinburgh*, vol. 1 (1788), p. 207. This paper was expanded and, under the same title, published as a book in 1795.

10. *Ibid.*, p. 214.

11. Thomas Hobbes, *Leviathan*, "Introduction."

12. Francis Bacon, *Novum organum* I, 81 (see also I, 129); René Descartes, *Discourse on Method* (1637), part VI. One may note, too, that for both Bacon and Descartes geological knowledge is from the beginning part of their new sciences. According to Bacon, his natural history goes beyond "nature free and at large" as it is present in "heavenly bodies, meteors, earth and sea, minerals, plants, animals" to focus "much more on nature under constraint and vexed"; it nevertheless explicitly does not reject the future possibility of forcing even Earth and sea out of their "natural state," as when they are "squeezed and moulded" to produce knowledge (Bacon, *Great Instauration*, "Plan," section 3). According to Descartes, "in place of that speculative philosophy which is taught in the Schools, we may discover a practical philosophy by means of which, know-

ing the force and the action of fire, water, air, the stars, heavens and all other bodies that surround us, as distinctly as we know the various crafts of our artisans, we can in the same manner employ them in all those uses for which they are suited, and thus render ourselves the masters and possessors of nature" (adapted from *The Philosophical Works of Descartes*, trans. Elizabeth S. Haldane and G.R.T. Ross [Cambridge: Cambridge University Press, 1931], vol. 1, p. 119).

13. Hutton, "A Theory of the Earth," *supra* note 9, p. 216.

14. *Ibid.*, p. 218.

15. Charles Lyell, *Principles of Geology: Being an Attempt to Explain the Former Changes of the Earth's Surface, by Reference to Causes Now in Operation*, Vol. I (London: John Murray, 1830), p. 1. Volume II appeared in 1832, volume III in 1833. All three volumes have been republished in facsimile editions and edited by Martin J.S. Rudwich, University of Chicago Press, 1991.

16. "The practical advantages already derived from [geology] have not been inconsiderable: but our generalizations are yet imperfect, and they who follow may be expected to reap the most valuable fruits of our labor" (*Ibid.*, Lyell, *Principles*, I, 4, p. 74).

17. Lyell compares progress in geology with progress in astronomy (*Ibid.*, *Principles* I, 4, pp. 73–74). Moreover, he argues, "the gradual progress of opinion concerning the succession of phenomena in remote eras resembles in singular manner that which accompanies the growing intelligence of every people, in regard to the economy of nature in modern times" in which "immaterial and supernatural agents" have been replaced by "fixed and invariable laws" (*Principles* I, 5, pp. 75–76).

18. See, e.g., David Kitts, *The Structure of Geology* (Dallas, TX: Southern Methodist University Press, 1977); Rachel Lauden, *From Mineralogy to Geology: The Foundations of a Science, 1650–1830* (Chicago: University of Chicago Press, 1987); H. LeGrand, *Drifting Continents and Shifting Theories* (New York: Cambridge University Press, 1988).

19. For Ernesto Mayz Vallenilla, *Fundamentos de la meta-tecnica* (Caracas: Monte Avila, 1990), meta-technologies do not simply enhance human capacities (in the way that the telescope and microscope, for instance, enhance human sight); they create new forms of sensing and action (such as radar and heat-seeking missiles).

20. See, e.g., Julian L. Simon, ed., *The State of Humanity* (Cambridge, MA: Blackwell, 1995), especially the Introduction and Conclusion.

21. Rachel Laudan, "Geology, Philosophy of," in Edward Craig, ed., *Routledge Encyclopedia of Philosophy* (New York: Routledge, 1998), vol. 4, pp. 25–28.

22. Thomas Kuhn, *The Structure of Scientific Revolutions* (Chicago: University of Chicago Press, 1962; 2nd ed., 1970).

23. As one indicator of this replacement, note that the *Encyclopaedia Britannica*, 14th edition (first published 1929), retained the term "geology" as the general term, while the 15th edition (titled the *New Encyclopaedia Britannica*, 1974) introduced the term "earth sciences." For a complementary perspective on the emergence of this term, see Gregory A. Good, ed., *Sciences of the Earth: An*

Encyclopedia of Events, People, and Phenomena, 2 vols. (New York: Garland, 1998), Introduction.

24. Rachel Carson, *Silent Spring* (Boston: Houghton & Mifflin, 1962). Apollo 8, December 21–27, 1968, the first circumnavigation of the Moon, produced pictures of Earth from space that quickly became icons of the environmental movement, which celebrated the first Earth Day a year and a half later. Consider, e.g., the following: The *Whole Earth Catalog* put an image of Earth from space on its Fall 1968 cover. *Life* magazine featured Apollo 8 coverage twice: "Discovery," Life, vol. 66, no. 1 (January 10, 1969), pp. 20–27, with a poem of the same title by James Dickey that ends: "And behold/The blue planet steeped in its dream/Of reality, its calculated vision stalking with/The only love"; and Anne Morrow Lindbergh's reflection on Apollo 8, "The Heron and the Astronaut: The Moon Voyagers and the Earthly Beauty That Beckons Them Back," *Life*, vol. 66, no. 8 (February 28, 1969), pp. 14–26, which concludes: "Through the eyes of the astronauts, we have seen more clearly than ever before this precious earth-essence that must be preserved. It might be given a new name borrowed from space language: 'Earth shine.'"

25. It is remarkable, for instance, the extent to which two standard histories of environmentalism—Roderick Nash's *Wilderness and the American Mind* (New Haven, CT: Yale University Press, 1967; 3rd ed., 1982); and Max Oelschaeger's *The Idea of Wilderness: From Prehistory to the Age of Ecology* (New Haven, CT: Yale University Press, 1991)–ignore, to mention but one example, the organic gardening movement.

26. K.S. Shrader-Frechette and E.D. McCoy, *Method in Ecology: Strategies for Conservation* (Cambridge: Cambridge University Press, 1993).

27. See, e.g., Lyell, *Principles* I, chap. 4, pp. 75 ff.; and III, chap. 26, pp. 383–384, *supra* note 15. For elaboration of this argument, see Andrew Dickson White, *A History of the Warfare of Science with Theology in Christendom* (New York: Appleton, 1896), especially chap. 5, "From Genesis to Geology."

28. James Mooney, "The Ghost-Dance Religion and the Sioux Outbreak of 1890," *Annual Report of the Bureau of [American] Ethnology*, vol. 14: 1892–1893, part 2 (Washington, DC: Government Printing Office, 1896), pp. 721, 724. A version of this quotation is cited by Mircea Eliade in *The Sacred and the Profane: The Nature of Religion*, trans. Willard R. Trask (New York: Harcourt, Brace and World, 1959), pp. 138 ff.; and in *Myths, Dreams, and Mysteries: The Encounter Between Contemporary Faiths and Archaic Realities*, trans. Philip Mairet (New York: Harper & Row, 1960), p. 155.

29. "Declaration on Soil" has been published in *Whole Earth Review, NPQ: New Perspectives Quarterly*, and other publications.

1 3

Sacred Earth

Karim Benammar

Karim Benammar brings a multicultural perspective to his work: Born in Algeria, raised in The Netherlands, he studied philosophy at the University of Sussex, England, Pennsylvania State University (Ph.D., 1994), and Kyoto University, Japan. Benammar is currently associate professor in the faculty of Cross-Cultural Studies at Kobe University, Japan. He has published articles on community, the relevance of myths, and the notions of excess and abundance in ecology.

Within the Earth sciences, questions concerning religion typically come up only within the context of debunking creationist claims that the Earth is 6,000 years old. Although these efforts serve a useful purpose, they also tempt us to ignore other, less antagonistic ways to describe the relation between the Earth sciences and the sacred. Whether it is in our backyards or in our national parks, our experience of the natural world is shot through with the sense of the numinous—the reaction of awe and wonder that lies at the basis of the religious impulse.

"Sacred Earth" argues that the standard account of our relation to nature has neglected this crucial aspect of our experience. Seeing the Earth, rocks, and rivers as sacred does not impede, but rather complements, a scientifically inquisitive and economically productive mind. Benammar argues that recent insights into the interconnectedness of all of nature, both organic and inorganic, have made the strict division between humans and nature obsolete. It is time to combine our scientific and economic views of nature with our lived response to the grandeur of the natural world. This sense of the numinous quality of nature is in fact manifested in the sense of curiosity and wonder that scientists bring to their work.

To travel the Earth is to be struck by its overwhelming size: the vastness of the outstretched plains, the depth of the oceans, and the endless skies. To walk upon the Earth is to be struck by the abundance of life everywhere—the deep forests, the smell of pine carried by the wind, the roaring of a waterfall, the incomprehensibility of every single snowflake being unique. Nature is wild, rough, untamed, overflowing, and boundless in its variety and fertility.

We have only recently come to realize how fragile the ecosystem that maintains us really is. A few degrees less in average temperature and we are heading toward another ice age; a few degrees more and we are subject to global warming. A small change in the chemical constitution of the air that we breathe, and our survival is at stake; a small change in the ozone layer that protects us and the sun that nourishes life turns into lethal radiation. We have come to realize that human civilization is an integral part of an interconnected ecological system. We have reached the end of the trope of undiscovered lands waiting for us, the limit of the boundlessness of the Earth.

Against this sense of endless vastness and limitless possibilities for growth, we are experiencing, for the first time, ecological limits to our growth.

Humanity is now a force powerful enough to change the geophysical makeup of the planet. We know that our mode of living is closely tied to the ecosystem on which it depends. How do we experience and understand our sense of belonging, of being part of a larger system that includes us and from which we cannot so easily distance ourselves? We are part of an ecosystem that is much larger than we are, which envelops us and flows through us, which we do not master and which affects us in a multitude of ways we cannot control. This ecosystem does not just include humans and animal life in all its forms; it is a system of human beings, winds, and currents, mountain ranges and snowstorms, underground rivers and burning deserts.

Broadly speaking, there are three ways in which human beings engage and are engaged by nature: a scientific way, an economic or utilitarian way, and what I will call a sacred approach. The scientific method probes nature, analyzes it, decomposes it into its constituent elements, and theorizes about it. The productive and economic approach assesses the use of nature for human benefit through such ventures as farming, exploration, mining, development, and landscaping. Against these profane conceptions of nature as something to be analyzed or exploited, however, there is also a long history of conceiving of nature as the place of the sacred. I will argue here that we need all three of these views of the Earth. While our culture has focused on the scientific and economic view of the Earth, we seem to have lost touch with the sense of the Earth as sacred. This sense of the sacred consists of a sense of belonging to the Earth and to the web of life, and of a sense of reverence and respect for life in all its forms and shapes.

13.1 THE SCIENTIFIC AND ECONOMIC VIEW OF THE EARTH

When we study the Earth scientifically, taking it as an object of our study, we analyze it, and we conceptually or actually decompose the stuff it is made of into smaller and smaller pieces. We trace the contours and geography of the Earth, chart its mountains and valleys, rivers and oceans, nooks and crannies. We plot and analyze ocean currents and cloud patterns, recording precipitation and average temperatures. We have become reasonably successful at short-term weather predictions. We also study the Earth by mapping its inner composition, and we probe it to reveal its structure. Through these various methodologies, we construct an image of the Earth as it is now, and reconstruct how it was at various times in geological history. This helps us to model the Earth's future contours and climate.

In all these activities we see the Earth as an object for our knowledge, which exercises our scientific curiosity. This curiosity is a beautiful thing, one of the more attractive qualities of the human species. However, through the scientific approach we are liable to become detached from the object of our study, seeing the Earth as an objective thing separate from us.

We also treat the Earth as a vast resource system to provide for human life. When humans were hunter-gatherers, they had little impact on the planet and left few traces. However, the impact of human habitation increased dramatically with the beginning of agriculture and urban civilization. Because of the rapid population growth and technological advances in the twentieth century, we now use so much of the Earth's resources that there is little left for the dwindling number of other species.[1] We have appropriated huge amounts of the land for agriculture, infrastructure, and housing. We clear-cut rain forests to make space for grazing cattle.

Nor is our use of the Earth limited to its surface. We probe the planet, mine it, extracting from it coal, oil, uranium, gold and other minerals, resources for our industry and commerce. Seeing the Earth in terms of its use-value, we cultivate it, landscape it, and reshape it to suit our pressing current needs. The economic or utilitarian approach views the Earth as one vast economic resource for the benefit of human beings.

The remarkable aspect about these two ways of considering the Earth, the scientific and the economic, is the way in which the Earth is set apart from the observer. The scientist observes and analyzes from an outside point of view, turning the Earth into an object of study and analysis. Economic and industrial approaches set themselves apart from the Earth by seeing it as a resource to be manipulated. Both the scientific and the economic or utilitarian ways are necessary and vital ways for human beings to relate to the Earth. But by themselves they cannot provide for a sense of belonging or of being at home in the world. This sense is found in taking the Earth to be a sacred place.

13.2 THE EARTH AS A SACRED PLACE

Against the attitude found in the two dominant modes of relating to the Earth in our time, the scientific and the economic, we can excavate a much older mode of relating to it, found in so-called primitive societies. In such societies, people held a deep sense of connection, in which the human world, the animal world, the vegetable world and the mineral world were all intertwined.

For traditional people, the Earth is something more than a material substrate: It shelters and nourishes them, and they return to it when their life is completed. In a word, the Earth is their home. As Joseph Campbell explains, in primitive societies all life was addressed with reverence and respect, producing a sense of the sacred:

> The Indians addressed all of life as a "thou"—the trees, the stones, everything. You can address anything as a "thou," and if you do it, you can feel the change in your psychology.[2]

The Earth was considered to be more than just an "it," to be studied or probed, bought and sold, landscaped, developed or mined. Considering the Earth as an "it" is tantamount to perceiving it as an object separated from our consciousness. Considering the Earth and all of life as a "thou" means recognizing its sanctity and being aware of the reciprocity and interconnectedness of human beings and nonhuman life.

Today many of us live in cities, shop in malls or in virtual space, fly halfway around the world for our holidays, and experience a large part of our reality through a television screen or computer terminal. Not surprisingly, we have lost most of this primordial sense of belonging to the Earth. This has entailed a loss of reverence, of the experience of addressing the natural world as a Thou. For many primitive people who lived and still live off the Earth, however, the whole natural world is sacred.

Although we tend to make a sharp distinction between organic life-forms and inorganic matter, most animistic religions consider there to be spirits everywhere and in everything. In Shinto, the original religion of Japan, not just trees and plants, but also rivers, waterfalls, Earth and rocks have spirits and carry spiritual energy. All aspects of life are considered to be sacred. According to the eighteenth-century Japanese scholar Motoori Norinaga,

> The word "Kami" [God or spirit] refers, in the most general sense, to all divine beings of heaven and Earth that appear in the classics. More particularly, the Kami are the spirits that abide in and are worshipped at the shrines. In principle, human beings, birds, animals, trees, plants, mountains, oceans—all may be Kami. According to ancient usage, whatever seemed strikingly impressive, possessed of the quality of excellence, or inspired a feeling of awe was called Kami.[3]

In Shinto, spirits dwell everywhere, in rivers, in mountains, in rocks, in all kinds of animals. Most Shinto rituals deal with establishing and continuing the proper relationships between people and the spirits of the mountains, forests, rivers, and animals. Countless shrines cover the rugged Japanese countryside: Consecrated places where spirits dwell are indicated and delineated by a thick rope garlanded with white paper tassels. Ancient trees, embodying a sense of time that transcends us, are singled out as places that are strikingly impressive and inspire a feeling of awe. Remarkable rock formations with particular shapes are understood to embody spirits. However, in many cases spirits are taken to dwell in nothing more than a patch of bare Earth.

The sense that there is consciousness in natural matter such as trees, plants, rock, water, or earth has become increasingly difficult for us to understand. Perhaps some of us can relate to the idea of animal consciousness, when we play with our pets and talk to them. We may even have a sense of vegetable consciousness when we find ourselves entranced by the calm majesty of a large tree or the joyous and vibrant aroma of a field. But how do we resonate with the consciousness of earth, rock, sand, or water? We might start from the feeling of awe when we are close to a powerful waterfall. We feel drawn in, resonate with it, feel ourselves to be somehow part of it. Imagine a river with its strong currents, or a stunning rock formation, or the rich soil of the rain forest, with its humid atmosphere, the powerful smell of the rotting logs and mushrooms, and the black mud.

A primordial sense of belonging was enacted by primitive societies by recognizing and celebrating spirits in trees, rocks, animals, wind, and breath. This sense of the sacred, of the Earth as the dwelling place of spirits, is found in most primitive cultures. For Australian Aborigines, the Earth's surface is sacred, and there is a strong connection between the geography of the outside world and the inner landscape of imagination and dreams. Aboriginal myths recount how the world was created by the Ancestors during the mythical time of the Dreaming, and how every striking feature of the vast Australian landscape is imbued with sacred energy. The sacred stories of the Dreamtime are connected to the landscape and preserved in mnemonic songlines that crisscross the Australian continent.

American Indians also celebrate the mountains and plains, the rivers and the air, as sacred. In the (possibly apocryphal) words of Chief Seattle:

> Every part of this Earth is sacred to my people. Every shining pine needle, every sandy shore, every mist in the dark woods, every meadow, every humming insect. All are holy in the memory and experience of our people.

We know the sap which courses through the trees as we know the blood that courses through our veins. We are part of the Earth and it is part of us. The perfumed flowers are our sisters. The bear, the deer, the great eagle, these are our brothers. The rocky crests, the juices in the meadow, the body heat of the pony, and man, all belong to the same family.

The shining water that moves in the streams and rivers is not just water, but the blood of our ancestors. If we sell you our land, you must remember that it is sacred. Each ghostly reflection in the clear waters of the lakes tells of events and memories in the life of my people. The water's murmur is the voice of my father's father.

The rivers are our brothers. They quench our thirst. They carry our canoes and feed our children. So you must give to the rivers the kindness you would give any brother.

If we sell you our land, remember that the air is precious to us, that the air shares its spirit with all the life it supports. The wind that gave our grandfather his first breath also receives his last sigh. The wind also gives our children the spirit of life. So if we sell you our land, you must keep it apart and sacred, as a place where man can go to taste the wind that is sweetened by meadow flowers.

Will you teach your children what we have taught our children? That the Earth is our mother? What befalls the Earth befalls all the sons of the Earth.

This we know: the Earth does not belong to man, man belongs to the Earth. All things are connected like the blood that unites us all. Man did not weave the web of life, he is merely a strand in it. Whatever he does to the web, he does to himself.[4]

In this description we see the main themes of a sense of the sacred belonging to the Earth. The animals and the trees, the mountains, rocks, rivers, and lakes, and also the air are taken to be alive, infused with spirits, and connected to human beings by invisible threads in the web of life. The knowledge that we are now rediscovering through our ecological science, through our studies of the effect of pollution on the world around us, through the realization of our interdependence with other species and with the natural world, is already present in these words.

This ancient wisdom is found in many stories and myths of primitive, traditional, and aboriginal people. In contemporary culture, myths are commonly understood as being little more than fairy tales. However, myths are better understood as the founding and regulative beliefs that guide a people. Myths embody the story of a people's relationship to the Earth, ranging from ancestor myths describing how the world came into being, to myths regulating the continuing rituals and observances for the preservation of harmony between the human and the natural world. Primitive people were related to the Earth by their sense of dwelling, by agriculture or by the hunt, but fundamentally by this sacred sense of belonging to the Earth.

The French anthropologist Lucien Levy-Bruhl has called this mysterious way of belonging to natural processes *participation mystique*, or mystical participation. It was fundamentally a self-understanding of human beings whose essence was to be connected to and involved with the greater scheme of things, the web of life. Because we have focused exclusively on our well-developed scientific and economic approaches to the Earth, we have lost this vital way of experiencing the world of which we are part. And through this loss we have also lost an understanding of ourselves.

13.3 RECONNECTING TO THE EARTH: A SENSE OF THE SACRED

During the Enlightenment, only human beings were believed to have souls, and animals were considered to be mere machines. Now, for some of us, our sense of consciousness has come to encompass animals and even plants, but we may still find it nonsensical to talk about consciousness in inorganic stuff such as soil, rocks, or water. In other cultures, however, as we have seen, people did not sharply distinguish between organic living beings and inorganic matter, between plants and rocks, or between animals and water.

If we reflect carefully on our own nature we will find that it is not so far-fetched to expand consciousness to include all matter. Rather, our present approach of restricting consciousness to what looks like and acts like us, our basic anthropocentrism, needlessly limits us and keeps us separated from the natural world. We can try to grasp this primordial sense of belonging by realizing that we are creatures, both organic and inorganic, constituted of water and minerals, made of the same stuff as mountains and rivers, as the Earth itself. The Earth nourishes us: Decaying matter and minerals in the soil allow plant life to grow, and water and sunshine feed life. The inorganic is what makes it possible for organic life to flourish. While food is essentially organic, either animal or vegetable, both air and water are inorganic, but they are vital building blocks of life. The water that human beings drink is not just the chemical compound H_2O, but also carries minerals and salts our body needs. We often forget that our body is constituted of more than 70 percent water, and that we react to the phases of the moon in ways similar to how the Earth itself reacts to the phases of the moon through the oceans' tides. From the perspective of animist societies, it is impossible not to feel that we are made of the same stuff as the Earth. And our scientific understanding also tells us unmistakably that we are part of nature, partly organic and partly inorganic, made of the stuff of rivers, oceans, and mountain ranges.

On a chemical level, the story becomes even more wondrous: The molecules of water we drink now we have drunk before, having passed through the rivers of the world, and through the bodies of countless other humans and animals. This remarkable fact illustrates our fundamental interconnectedness with both organic and inorganic substances.

The cycle of breath allows all life to grow. We breathe in what plants breathe out, and they breathe in what we breathe out, maintaining a mutually sustaining relation. We now know that the forests are the lungs of the world. For many Eastern religions, the breath is what connects the inner with the outer, the inside with the outside. The inner rivers of our lungs, the inner geography of our veins and arteries are connected through the breath with the outer geography of air currents. Breathing is a continuous exchange between our bodies and the outside world. In meditation practice, the regulation of the breath allows us to become calm, to attune ourselves to the world around us, to dissolve into it.

The air we breathe is a gaseous substance, a fine balance of chemicals; if the air is polluted, we develop rashes, sores, and cancers. If the admixture of chemicals in the air is perturbed, we may die, asphyxiated. David Abram proposes that instead of seeing the air as emptiness, we see it as a living presence that surrounds us, but that is also in us, that fills us, that embodies our connection to the elemental forces of the planet:

As the experiential source of both psyche and spirit, it would seem that the air was once felt to be the very matter of awareness, the subtle body of the mind. And hence that awareness, far from being experienced as a quality that distinguishes humans from the rest of nature, was originally felt as that which invisibly joined human beings to the other animals and to the plants, to the forests and to the mountains.[5]

I have tried to bring us back to the awareness that human life is composed of animal, vegetable, and mineral life by pointing to what we are made of physically. Our human nature, quite literally, consists in being part water, stone, and air, and part organic animal cells. We are the Earth that we walk upon, the rock we climb, the tree we lean against, and the flower we hold. Having a sacred attitude toward the Earth is also having a sacred attitude toward oneself—respecting the Earth as a holy place is respecting oneself as a holy place.

How can we regain the sense of belonging to the Earth, the sense of being part of the web of life, the sense of oneness with nature that surrounds us, that lives in us and through us? Myths and ancient animistic religions are relevant today because they speak to us about the sanctity of the Earth. Primitive people had a sense of belonging that we have lost: They understood the spirits in waterfalls and in the wind, the communication between animals and human beings, and the intimate connection between the waking world and the dream world. Primitive people understood the sanctity of the Earth—they lived the experience of human life as only one part of the web of life. We have rediscovered this web of life in our contemporary ecology studies and call it an interrelated and integrated ecosystem.

The sense of the sacred had been marginalized in our human-made environment of cities, suburbs, and freeways. In our culture, consecrated places are limited to churches or temples. However, one avenue for rediscovering the natural world as a sacred place is through the feeling of awe and reverence we experience when we are in nature.[6] According to the Shinto religion, the presence of spirits is felt most strongly in impressive places, which inspire awe in human beings. This is true of our wild spaces, which we could call natural temples. When we go to Yosemite, the immensity and arresting beauty of its massive cliffs overwhelm us. The size and age of the Sequoias dwarf us. Similarly, seeing the interplay of clouds and rain on the surface of the sea rekindles some long-lost notion of the infinite variety and bounty of the natural world. Some of the Earth's elemental forces become evident to us. These stunning natural places dwarf our mental constructions just as weather patterns, hurricanes, and earthquakes mock and tear through our skyscrapers and bridges. The elemental forces that were so frightening to our ancestors can still overwhelm us.

When we go to these vast places it is to feel the strength of the elemental power of rock formations and old-growth forests, or to revel in the excessive colors of corals teeming with life. These are impressive places, which inspire awe in us, places where spirit is concentrated, where its effect on us is strongest. Spirit is also present in the concrete of cities, in the stone we have quarried and ground, in our highways, in the steel of our automobiles, and in the chopped-down wood of our houses and furniture. However, it is much easier for the human mind to be struck and displaced by the immensity of the rocks, or for the spirit to soar in the vast empty spaces over the sea. It is easier for the mind to reconnect to the web of life by being surrounded by the elemental, away from the clock-time of city consciousness and the digitizing streams of television and computer that command so much of our attention. It is easier for the body to be reawakened to its primordial sense of belonging in nature than in our civilized lifestyles in which we have minimized the effects of temperature, seasons, day and night, dampened all the natural cycles of which we are part.

The Redwood forests, the Grand Canyon, and the Great Barrier Reef have become places of pilgrimage. We can feel that rivers and waterfalls are alive, appreciate the stillness of a lake and the roughness of the waves on the sea. Water reminds us of the changing flows inside of us; water flows through rivers as blood flows through our veins. We can get a sense of a time span radically different from our own when we are close to a 700-year-old tree—a life span that dwarfs ours. We can sense how stones embody time, realize that the mountains and Earth we walk on represent petrified time, our own prehistory. Rock formations have slowly accumulated over hundreds of thousands of years, formed through silt deposits from oceans at the dawn of time, or forged with unimaginable violence through volcanic eruptions, created by molten excretions from the Earth's fiery core. Rock formations embody a time that is beyond our comprehension: even though we can express this time in numbers, we cannot grasp it in terms of lived experience. And yet these rocks, too, are alive in some fashion that exceeds and transcends our experience and understanding. As Jim Cheney observed:

> Although the very idea that we might have ethical relationships with *rocks* is nearly unintelligible to the Western mind (or proof of mental imbalance), many American Indian cultures take rocks quite seriously, often regarding them as the oldest and (in some sense) wisest of beings.[7]

These impressive natural places, these sacred places, are the places where we return from our exile, from our wandering in abstraction, where we come home to a more primordial sense of who we are. We return to a sense of self surrounded by mountains, ferns, moss, animals, surrounded by spirits, in places where these spirits are strong and wild. When we are in the wilderness, we are reawakened, we can come home to the sacred Earth. These are still the sacred places where, as Chief Seattle tell us, "man can go to taste the wind that is sweetened by the meadow flowers."

The beauty of the scientific way of looking at the world is the sense of curiosity involved in making sense of things, in taking them apart, in making hypotheses and in running experiments. Recently, science has also shown the limitations of our understanding of the world through ideal abstractions. Chaos theory teaches us how mapping the whirls of an eddy in a river or the shapes of a cloud of smoke requires a whole new way of conceptualizing the problem. Our scientific understanding is being brought to a more humble attitude toward the richness of natural processes.

The Earth sciences study climate patterns, the history embedded in silts and deposits, the location of minerals and ores, and the limits of our water resources. The study of the Earth's geology and geography, the oceans and the air, should be one of the fundamental practices through which we can reconnect to the sense of the Earth as a sacred place. Geologists are by their very nature intimately connected to the Earth, because they study it and because their work often involves assessing the use-value of the Earth as a natural resource. Earth scientists also spend much time in nature, among rocks; they are connected closely to the Earth, and especially to the inorganic part of the

Earth that seems so alien to our human consciousness. For these reasons, Earth scientists are in an excellent position to help the population at large gain mnan awareness of the Earth as a sacred place. One of the tasks of the Earth sciences should be to contribute to understanding our place on Earth, to help us recognize the value of the inorganic in the web of life.

Although conflicts often emerge between a profane and a sacred conception of nature, an ecological view of nature must include a spiritual component. This is also true for the scientist: When physicists speak of understanding the mind of God through their discovery of physical laws, they are alluding to a conception that transcends a purely mechanistic view of nature. Geologists should cultivate a similar attitude: Considering the Earth, rocks, and rivers as sacred, respecting this sanctity and having a sense of humility toward nature does not impede but rather complements a scientifically inquisitive and productive mind.

FURTHER READINGS

For an excellent introduction to myths and their relevance to the human mind and to contemporary life, see the famous interview series of Joseph Campbell conducted by Bill Moyers and published as *The Power of Myth* (Doubleday/Anchor Books). It is also available as a video of the six-part PBS television series.

An Invitation to Environmental Philosophy contains many good articles. The collection *Ecopsychology* contains essays dealing with the relationship between the environment and the various psychological needs and desires of human beings.

NOTES

1. Humans directly use 40 percent of all of Earth's land resources. See Norman Myers and Julian L. Simon, *Scarcity or Abundance? A Debate on the Environment* (New York: W.W. Norton, 1994), p. 93.

2. Joseph Campbell, with Bill Moyers, *The Power of Myth* (New York: Doubleday/Anchor Books, 1988), p. 99.

3. Quoted in Kenji Ueda, "Shinto," in *Religion in Japanese Culture* (Tokyo: Kodansha International, 1996), p. 35.

4. Chief Seattle, quoted in Campbell, *op. cit.*, pp. 42–43.

5. David Abram, *The Spell of the Sensuous*, (New York: Pantheon, 1996), p. 238.

6. See Robert Greenway, "The Wilderness Effect and Ecopsychology," in Theodore Roszak et al., eds., *Ecopsychology* (San Francisco: Sierra Club Books, 1996).

7. Jim Cheney, "The Journey Home," in Anthony Weston, ed., *An Invitation to Environmental Philosophy* (Oxford: Oxford University Press, 1999), p. 144.

14

Ecological Emotions

Alphonso Lingis

Alphonso Lingis is Professor of Philosophy at Penn State University. A leading scholar within the Continental tradition of contemporary philosophy, Lingis's work is distinguished by the juxtaposition of academic philosophy with extreme events of life. Years spent living and writing in third world countries has resulted in a unique style of thought where contrasts of wealth and poverty, personal and cultural dislocations, and intense, transitory relationships between people living radically different lives illuminate and are illuminated by classic texts in the history of philosophy. Lingis's books include Excesses: Eros and Culture *(SUNY Press, 1983) and* The Imperative *(Indiana University Press, 1998).*

This essay offers an account of our emotional responses to the natural world. The phenomenological tradition within contemporary philosophy argues that our interaction with things is not in the first instance scientific; rather, our various investigations are given direction by the demands of our affective life. Emotions are intelligent responses to the world that guide all subsequent projects, scientific or otherwise.

Lingis provides a wide-ranging analysis of our emotional connection to nature, organized under the headings of respect and Promethean fear. He then turns to how the elemental forces of earth and sky, oceans and vegetation, mountains and canyons affect us. In ways reminiscent of the work of the French philosopher Gaston Bachelard, each of these elements is described as eliciting distinctive emotional reactions that shape the directions our lives take. Lingis's account highlights the ways in which a culture's affective responses to nature fundamentally determine its projects.

All that is made by humans is, directly or indirectly, derived from natural materials, and all structures built by humans eventually disintegrate through the effects of natural forces. Nature is contrasted with culture. But we humans have now occupied the entire planet, made it our habitat. Where do we encounter nature?

Is not the perception of nature a perception of cultured, cultivated, or culturally delimited and preserved nature? We see trees and grass in our gardens, crops in our fields; even the weeds in those gardens and crops are imported plants or plants that have become invasive because of the opportunities cultivation of the soil provides them. A nature park is not nature, but a park; a wilderness that is preserved is a delimited wilderness, an ecosystem that will therefore sustain fewer species. Is it not through our use of the land, transformation of mineral deposits and petroleum, and production of consumer goods that we envision the natural as what is there by itself and grows of itself? Is it not in the disintegration of what we use, transform, and produce—the tarnishing of our silver, the metal fatigue of our jet airplanes, the crumbling of the stones of the Pyramids—that we envision the natural as what is indifferent to our personal and cultural projects? Is it not through our efforts to understand what is there by itself and grows of itself and is indifferent to and recalcitrant to our projects, through animist or deist concepts or through concepts of determinist empirical laws, that outside nature exists for us?

It is a commonplace to contrast the Judeo-Christian notion of nature as something to be conquered with the concepts in Buddhism, Hinduism, and "animist" religions. The concepts of obedience to the laws of nature, of harmony with nature, as well as the concept of involvement in an ecosystem are all specific to cultures. Concern for nature today is theoretically shaped by the new ecological sciences and practically focused by worries about environmental pollution and destruction. What is understood by nature is shaped by the concepts in use in these theoretical disciplines and practical assessments.

14.1 RESPECT

Our sciences do not only construct a representation of nature and our practical undertakings do not only transform nature; they motivate specific emotions. Environmental worries motivate restraint before the unmapped and uncontrolled; empirical sciences enjoin a sentiment of respect for nature.

Respect, according to Immanuel Kant, is "something like fear, something like inclination." It is an intellectual feeling, a feeling induced not by sensuous impressions but by understanding. Kant identified as the primary meaning of respect not respect for a sovereign, but respect for law. Respect for the sovereign is not only fear of his power and awe before his splendor, but recognition of the universal validity and the necessity of his decrees.

Respect—in Kant's German, *Achtung*—involves a kind of perception: an attentiveness to the space of another. But respect is not simply a mental attitude; it is something practiced. To respect another is to not encroach upon the space of another. It is to allow another to have his or her space. It involves a sense of another as a being unto himself or herself or itself.

Respect marks a limit to our action; it is an inner inhibition. But to respect is also, positively, to esteem, to value. It is to defer to. To respect someone, or something, is not only to inhibit our own invasive impulses before him or her or it, but to some extent to subordinate our behavior to his or hers or its.

The contemporary sentiment of respect for nature has the specific meaning of respect for the laws of nature—acknowledgment and acquiescence before the known laws, restraint before unknown laws. For what is practically experienced as beneficent and as destructive and what is theoretically experienced as unpredictable and chaotic is taken to be the domain of laws as yet unknown.

Hence the paradox: that, while it is through use, transformation, and production that we encounter and understand the natural as what is there by itself and grows of itself, and through protection, repair, and renewal that we encounter and understand the natural as what is indifferent to our personal and cultural projects. Our understanding generates the sentiment of respect that enjoins practical restraint before nature.

14.2 PROMETHEAN FEAR

But to identify respect as the emotion that the sense of nature awakens in us is to identify ourselves with our practical undertakings and our theoretical efforts. Is not to do so to identify ourselves as cultural agents? Is it not, paradoxically, to ignore our own nature, our existence as natural beings? Are there not emotions that arise in relationship to nature that are not intellectual emotions, not emotions induced by understanding of empirical laws?

Bernard Williams identifies two basic kinds of emotional relations to nature: gratitude and a sense of peace, on the one hand, and terror and stimulation on the other.[1] The expanse and the forces of nature about us disturb us, trouble us, excite us, and terrify us. There was the fear of the first humans before the perceptible dangers of their animate and inanimate environment, and our fear still of diseases, floods, and tornadoes. So much of our activity is provoked by weeds and drought that threaten our food supply, storms that threaten our homes, climate changes that threaten our health. Tranquillity, assurance, and bliss are not only states we tend toward naturally, states natural to us; they are states of concord and repose in our natural environment. Even if they are achieved by the most artificial of means—a totally controlled environment, psychotropic drugs—they echo emotional states induced by natural settings. Tranquillity, security, and bliss in states in which we feel gratified, and, because they require a certain presence of our environment, states that induce gratitude.

These two named emotional poles—a sense of peace and gratitude, and of stimulation and terror—in fact designate a panoply of most diverse emotions. The first pole covers the restorative effects of a weekend spent walking the autumn forests, the trance state induced by watching a butterfly emerging from a chrysalis, the melancholy consolation of a tropical sunset, the euphoria the sailor feels when he is back on the open seas—and much more. The second pole covers the sense of contingency one feels in the midst of everything made by humans, whether it be a still-used Roman or Inca road or a new suburb, the sense of vulnerability one senses in multibillion-dollar jet airplanes and in a new-built condominium when one insures it against "acts of God," the nervousness one feels upon reports of earthquakes or outbreaks of mutant viruses, the anxiousness with which one waits for the physician to assure a parent that one's infant shows no sign of genetic defect, and the malaise with which one lets one's mind follow the astronomers who chart the rate of extinction of stars and galaxies.

These "basic kinds of emotional relations with nature" contrast with the emotions induced by the current conceptual representations of nature. The biblical picture of nature as a garden planted and stocked with animal species for humans to name and use, and the technological representation of nature as a fund of materials and energy open to calculative thought and technological transformation, sanction a predator emotion of confident dominion. Environmentalists and ecologists promote a sense of intimate involvement and community. Neopagan religions arouse sentiments of benevolence and altruism.

The emotions of stimulation and terror—the anxiety that we feel before blizzards and floods and that troubles our minds when we read of an immense meteorite hitting the planet Jupiter—expose something beyond or behind the biblical picture of the environment as a garden planted for human use, and beyond or behind the technological representation of nature as a stock of raw materials and energy sources—something which is *nature*. The wonder upon seeing sprays of coral fish in the surge of tropical waters, and upon seeing just three days after, small plants creep out of the ashes of the thermonuclear-devastated Hiroshima, attaches to *nature*.

This twofold emotional contact with nature found expression in Western art when aesthetic theory distinguished the sense of the beautiful from the sense of the sublime. From the end of the eighteenth century, in contraposition to the prospect of technological control of all of the planet's resources, plastic art devoted itself to nature in its gratuitous harmony but also its awesome and terrifying visages.

It is true that Immanuel Kant's aesthetics of the beautiful and the sublime did not affirm the separateness and fearfulness of nature. For him, the vision of beauty, a beauty of harmonious forms, induces dispassionate composure in the observer. And the vision of the indefinitely irregular forms of cliffs and of the immeasurable forces of storms at sea induces in the unendangered viewer the idea of infinity, with which the mind rises over every perception of the indefinite and the unencompassable.

In contrast, our contemporary plastic art finds beauty in jagged, ragged, indecisive forms, and presents what is not representational and does not induce any ideas, does not exalt the intellect with the idea of infinity but stupefies it with the sublimity of turbulent and incendiary sensuous forms and forces. This art has lost its cultural innocence, and rediscovers a sublimity that is inhuman and ominous.

It is facile to show that the environmentalists and ecologists who promote a sense of intimate involvement with nature, and the neopagan religions who promote a sense of benevolence and altruism toward all species with which we share the planet, overlook the innumerable species that feed on their own eggs or offspring, overlook the predator species, overlook the natural extinction of species. And they overlook the omnivorous, hence predator, nature of the human species from its first appearance. Certainly, it can be argued that humans are uniquely able to detach themselves from predator behavior; we began to do that as soon as we domesticated plants and animals, and for millennia the great majority of humans have been most of the time vegetarians out of necessity and a significant number out of choice. And certainly it can be argued that humans are uniquely able to understand that our future survival depends on detaching ourselves from our biblical and technological attitudes of confident dominion over the planet.

But this detachment is first in our basic emotional relations to nature. The perplexity we feel before crops dying of some disease and the terror when our jet airplane pitches in turbulence revive in us the ancient sense of our vulnerability in an indifferent and dangerous nature. The sense of assurance, peace, repose we feel before a brook sparkling at the bottom of a canyon, a mountainscape blanketed in new-fallen snow, flows into a primary sense of gratitude, for it contains a sense of something inexplicably there, simply given, there by chance and good luck. In this sense of a nourishing and halcyon setting given by good luck trembles a fear that we may ourselves destroy it. In these emotions we feel our culture's distinctness from nature.

Our sense of returning to a natural assurance and peace, and the inherent sense of gratitude that stirs there, motivate restraint with regard to nature, as does our sense of vulnerability and our fear before natural and uncontrolled forces. This sense of restraint is not simply animated by the intellectual sentiment of respect. Restraint in the face of nature is produced by fear—a fear not just of the power of nature itself, but, Bernard Williams writes, *Promethean fear*, a fear of taking too lightly or inconsiderately our relations to nature.[2]

14.3 TORMENT

Respect is only an intellectual sentiment; it is also essentially negative. To respect another is to allow another to have his or her space, his or her practical space, and his or her fantasy space; it is to not encroach upon the space of another. It is to acknowledge that people are an end in themselves by not treating them as resources or a means. To defer to a person is to inhibit our own invasive impulses before the person. Is not respect for another a noninvolvement and noncommunication with the person?

Another person is something perceived, exposed to us, and circumscribed by our vision. His intentions are exposed by the way he stands and moves, by his body kinesics, and by his words. His actions and his body are also exposed to our operations; his space is in the common world. He is exposed to being violated, outraged, wounded by us (when we pay for high-tech surgery, doctors are lured from poor regions; when we sit down to a gourmet restaurant meal, land is being diverted from basic food crops in remote lands). And for us to defer to another, to subordinate our behavior to some extent to him, is to expose ourselves to him—expose ourselves to being violated, outraged, wounded by him. To approach another with respect is to expose our seriousness of purpose to the flash-fires of his laughter, to expose our cheerfulness to the darkness of his grief, to let him put his blessing on our discomfiture and suffering, to expose ourselves to the shock waves of his curses. It is through our wounds that we communicate.

The respect shown for nature is a sentiment of acquiescence to the known laws of nature, and a passivity before the known. Is not *Promethean fear* also a reflexive and essentially negative sentiment? Must not the practical restraint before nature, urgently enjoined by our environmental and ecological worries, issue instead from the most intense contact with nature—and therefore from the most vehement positive and carnal emotions? Our emotions are movements outward; they are not simply the recoils of impact of what struck us or rumblings of contentment over content absorbed. Our strongest and most intense emotions plunge further into the depths of the landscape, the skies, the oceans, and the earth. "There exists…an affinity between, on the one hand, the absence of worry, generosity, the need to defy death, tumultuous love, sensitive naivety; on the other hand, the will to become the prey of the unknown. As far as they are able (it is a quantitative matter of strength) men seek out the greatest losses and the greatest dangers…. If a good measure of strength does fall to them they immediately want to spend themselves and lay themselves bare to danger. Anyone with the strength and the means is continually spending and endangering himself."[3] To fear nature is to be decomposed by that fear, and to know the bliss of concord and repose in the bosom of nature is not to be gratified and consolidated but to be emptied of ourselves. Is it not in the most intense exhilaration and the most intense laceration—through titanic and not Promethean emotions—that we come to know what is, what happens, in nature?

14.4 EMOTIONS OF THE VEGETATIVE BODY

When our moralists speak of living according to nature and respecting nature and our own nature, it is the spectacle of flourishing fields and forests, rising upright from the Earth to the sun, that they invoke. They are not thinking of the sludge oozing in swamps where nature asserts its indomitable power in decomposition, teeming bacteria, snails in whose bodies are swarming Bilharzia protoctists, and mosquitos in whose bodies are proliferating malarial plasmodia. It is flourishing fields and forests that the ancient stoics, Emerson and Thoreau, Nietzsche and Hegel, the evolutionists with their tree or multistemmed bush of evolution and the ecologists of today invoke as the natural landscape of human rectitude and justice. Our subspecies of primate, standing erect, prizes the upright, equivalent for us to dignity; prizes rectitude, equivalent for us to righteousness; prizes the elevated; prizes eyes turned to the skies and the heavens, for us equivalent to the decent, the noble, the ideal, the sacred.

The branches and the twigs of the plants of meadows and of the linden trees and great elms presiding over the main streets of our towns give us the vision of an architectural order, an ordered distribution of each part in its own place, post, and function—a compelling everyday vision of justice and harmony. Each in its place the individual leafy stems exhibit purpose and dignity.

The sequoias are celebrated as the most noble of trees; rising as great upright poles 250 feet into the sky, with only the sparsest of branches, they are the one thing in nature that is sacred to all Americans. Their endurance, they that see so many peoples and societies and regimes come and go in America in the course of the 2,500 years they live, materializes fortitude and steadfastness for us. We tolerate, with bad conscience, the commercialization of our artists, our women, even our churches, but shrink back from cutting down the giant sequoias for timber. As long as they stand tall, there is still something noble and just in America.

But fields and forests elicit other kinds of emotions. Unseen below the uprightness of fields and forests rising to the sun, roots descend, wind and knot and wallow like worms in the wet earth, lured by rank decomposition. They, however, must not be seen. Roots turned up in the air and sun die and the whole plant dies with them; floods and storms that uproot the fields and the forests are disasters. The sequoias are not the oldest monuments in nature; in Africa and Australia there are baobab trees that are also 2,500 years old. They have thick, squat trunks, and then a brushy tangle of branches that look like the roots of trees uprooted by storms, and with hardly any leaves. They are the very image of what is inverted, perverted in darkest Africa. Our visceral sense of the base, the low, the mean, the vile in certain behaviors, attitudes, and inclinations designates their affinity with the dark and dank earth, with rot and decay.[4]

Some roots are particularly indecent. The mandrake root, short, thick, white, and bulbous, often branching into two at the bottom and often branching twice, presents an obscene image of a naked human body, headless and without muscle delineation. Carrots look too much like human penises, turnips like swollen testicles. These images give us the sense that there are base roots in our bodies—our penises, our genital organs, our naked fingers and toes on collapsed bodies pushing in disordered directions in sweat and darkness. To expose in the air and sunlight these roots pushing into slime and darkness, this crawling, squirming decomposing body, would be a moral disaster.[5] But there are emotions that open tumultuously upon these base organs in our bodies and upon all subterrestrial nature.

As our bodies become orgasmic, our posture, held oriented for tasks, collapses, the diagrams for manipulations and operations dissolve from our legs and hands, which roll about dismembered, exposed to the touch and tongue of another, moved by another. Our lips loosen, soften, glisten with saliva, lose the train of sentences; our throats issue babble, giggling, moans, and sighs. Our sense of ourselves, our self-respect shaped in fulfilling a function in the social environment, our dignity maintained in multiple confrontations, collaborations, and demands, dissolve; the ego loses its focus as center of evaluations, decisions, and initiatives. The sighs and moans of another pulse through our nervous excitability; spasms of pleasure and torment irradiate across the nonprehensile surfaces of our bodies, our cheeks, our bellies, our thighs. Our muscular and vertebrate bodies soften and melt, transubstantiate in mammalian sweat and reptilian secretions and releases of hot moist breath nourishing the floating microorganisms of the night air. The extreme pleasure of orgasm, the most intense pleasure we can know, is borne by an anguished abandon to cadaveresque decomposition. In the anxiety and exhilaration of orgasm, convulsion of what the ancients called our "vegetative soul," we sink into the living depth of our nature and into the depths of vegetative nature.

14.5 CELESTIAL EMOTIONS

Along the Pacific coast of Peru is a narrow strip of land walled in on one side by the Andes and exposed on the other to the ocean, kept cold all year by the Humboldt Current coming up from Antarctica. The air over this coastal strip, heated by the sun, expands out over the cold ocean so that moisture from the ocean does not drift back over the land. It never rains; most of this coastal plain is drier than the Sahara. Halfway down this coastal region, on the flat pebble desert of Nazca, there are great splays of shallow lines inscribed over 200 kilometers of the surface. The terrain is so flat and the lines so long that no one standing there could make out the patterns—not even those who made them; it was when Maria Reich in the 1930s, using surveying equipment, transcribed them on graph paper that the patterns and diagrams could be seen. Archaeologists have determined that they were made over a thousand-year period by a people who vanished a thousand years ago. A few lines have been coordinated with points of the horizon when the sun, and certain major stars, lie at the equinoxes. The scholars working on the Nazca lines believe they must be some kind of cosmic diagram—but what they depict, and for whom, remains incomprehensible.

Five hundred miles south of Nazca stands the colonial city of Arequipa. From there it is a four-and-a-half-hour van ride on the dirt road around the volcanos Misti, Chiacoan, and Pichi-Pichi. The terrain is all volcanic tufa, white swirls that seem to be poured from a bowl, and so arid it is almost completely devoid of vegetation.

At length you reached the brink of a fault in the continental tectonic plates: the Colca Canyon. Far below a meandering river has dug it to depths deeper than any other canyon on the planet. A path descends to the flank of the canyon. Here and there at the bottom of the canyon there are some Indian hamlets. These people were passed over by the Inca empire and are too alien and too poor to be enlisted in the Peruvian state today. You were able to find lodging for the night. The full moon shone in a splendor seen nowhere else; looking at it between the narrow walls of the canyon was almost like looking at it through a tube. Its crystal body and its light on the glaciers high over the canyon made the night numinous. You walked down the lanes, sensing the discrete but trusting proximity of the dark forms of strangers. You ended up in a room where four young men were playing a llama-hide drum, reed pipes, and two stringed gourds. The music had wild rhythms that pulled you up from your seat to dance and tragic melodies that tore your heart out. A young man knew enough Spanish to propose guiding you in the morning.

You set out at four o'clock. The natural terrain, from the glaciers down sheer bare rock and then to the slopes created by erosion and down into the deep gorge cut by the river, was a display of ever different cuts, different colors of rocks and cliffs. But equally astonishing was the terracing, begun at least 1500 years ago, which had faceted all the flanks of the canyon from the sides of the river up hundreds of feet. There were thousands of miles of stone walls supporting these terraces, forming interlacing amphitheaters from the river to the heights of bare rock and the glaciers above. The terraces were growing queñua, potatoes (originally brought to the outside world from Peru) and many species of high-altitude corn and were irrigated by the melting glaciers and the springs and cascades that release the groundwater. The Colca Canyon has been home to humans who created for themselves an advanced culture and worked the gorge with the patience and vision of a diamond-cutter. The women wore long, multi-colored bouffant skirts and broad-rimmed hats; as the canyon turned around mountains you noticed that their hat-style changed in different settlements. You had read that two different populations settled the canyon, revering two different sacred *apus*, mountains, and they reflected the forms of their respective sacred mountains on their own heads, placing molds on their infants' skulls so that, in the one group, their skulls would be flat, and in the others, high and cylindrical. When the missionaries grouped the people into fortified villages called *reducciones*, they forbade skull-molding of children, and imposed Spanish dress. The one group then devised high and round hats, the other flat hats.

On both sides the mountains were glacier-covered, but volcano Mt. Sabancaya was gushing out dense clouds of smoke. Earthquakes shook down a village at a place called Maca two years earlier, and now the quakes were starting up again at the same spot; you went down to look at a big crack that, the local people told you, had just opened up the night before. Your slow and wearing ascent up the canyon, the grit crumbling under your shoes, measured the time of your effort and of your life against the geological epochs of the planet's crust and rock layers. The scale of geological time diminished you and destines any footprints you leave here—indeed, any works you build in your lifetime—to erosion and mineral decomposition.

You trekked to the top of the terracing, and, chewing coca leaves, climbed ever higher; finally you reach the Condor's Cross, where the canyon is deepest. You had read that the Grand Canyon of the Colorado is 1,638 meters at its deepest, that here the Colca Canyon is 4,174 meters deep. It is also very narrow, a knife-cut through 12,000 feet of granite, at the bottom of which like a crinkle of mercury you see the river. All around, the glaciers of the Andes begin to blaze with the rising sun. Loftiest is the volcano Mt. Mismi, whose melting glacier is the source of the Amazon. Billowing in the sky are the sulfurous fumes of the Sabancaya volcano you passed hours earlier in the dark. You settle on a boulder in this uninhabitable mountainscape. No human enterprise could take hold here; you could form no project here, not even an exploratory hike. Even if you had any shreds of vocabulary of his language or he of yours, you would not have anything to say to your Indian companion, not even any question to ask him. The discursive movements of the mind, staking out paths, laying out positions and counterpositions, are silenced, deadened. As soon as the sun has emerged over the peaks of the Andes, it turns the whole cloudless sky magnesium-white. Its radiation spreads over your face and hands like warmth, although the thin air your heaving lungs are pumping in is cold. After a long time spent motionless, you are aware the sun is now high in the sky.

And then, well before seeing it, you are aware of the condor, like a silent drum-roll in the skies over the glaciers. Your eyes are pulled to a speck taking form in the empty radiance, imperceptibly becoming bigger, becoming a great bird never once flapping or even shifting its wings, soaring down from a great height and then into the canyon, descending to eye level in front of you before gradually descending deeper and becoming lost to sight. It is the first condor you have seen, with its 15-foot wingspan the largest flying bird on the planet. This one is brown, a young female, sur-veying the desolate cliffs and avalanches for carrion. And then—an hour, two hours later?—there are two: Again you know they are there well before they are visible. They are soaring close to one another, circling companionably in the airless heights. When they are overhead, you try to gauge their height, judging they are above you halfway again the depth of the canyon—that is, some 18,000 feet. You, who could hardly climb much higher than your present 13,000 feet, feel your eyes, your craving, your fascina-tion plunging to their almost immaterial realm, falling up into the region of death.

You are nothing but a vision, a longing, a euphoric outflow of life hanging onto the flight of the condors. Their flight comes from a past without memory and slides into a future without anticipation. You are cut loose, unanchored, without guy wires, drift-ing in the void of the sky. You know nothing but the flight of the condors, feel nothing but the thin icy air, see nothing but the summits and ice cliffs of the Andes and the granite walls of the canyon below your flight. You are alive to nothing but their bodies and their soaring, you are alive for nothing but for them.

14.6 OCEANIC EMOTIONS

Climbing the rocky coves and promontories of Sydney Harbor, hoping to catch sight of whales, mocked by the swift slippery seals, one gets tired. A hundred-sixty pounds of mostly salty brine in an unshapely sack of skin—one's legs fold and strain to hoist this weight; the joints ache. One can certainly understand the dolphins and whales, mam-mals who evolved on land, but long ago returned to the ocean. Still, movement on the Earth's surface is not simply a blunder on the part of evolution, as the case of serpents and cheetahs show. Still, one can also return to the ocean, with the whales and seals, strapping on an air tank and fins.

At closing time, I waited at the Manley Oceanarium. The staff had given me leave to enter the shark tank after hours with scuba gear. If walking across the surface carry-ing my body weight on jointed legs was awkward enough, doddering across the base-ment of the Oceanarium with buoyancy compensator, air tank, regulator and gauges, eye mask, weight belts, and flippered feet was ludicrous. But once inside the tank, having achieved buoyancy, I could slide through the water with my rubber fins. I gaped in wonder at how different the light and colors of the water were from the way they looked from the tunnel in the tank where visitors view them in the regular opening hours. Of course, it is still diving in an artificial reef. But the Oceanarium is so big (the tank holds 4.6 million liters, and they do not know how many animals are in there— thousands) that the behavior of the animals is not different from their behavior in the ocean. And in a whole lifetime of diving the oceans how often could one get that close to great sharks? I was told to avoid them bumping into me, and not to touch them, for sharks, like all fish, have a slime coating on their skin that protects them from bacterial infections and that my touch could break. For an hour I watched huge sharks passing inches from my eyes, great rays folding their bodies over my head.

It is the cartilage, not bones, they have that gives them that extraordinary suppleness of movement. Their sleek bodies are as hydrodynamic as our bodies holding themselves up off the ground are unaerodynamic. The rays are classified by science as belonging to the same family, but their movement is totally different from that of sharks. They are disks that shimmer and glide. Our bodies walk in a succession of falls stopped by a lurching of bones; their whole bodies are musically rippling. Their movement is disinterested, movement as the nature of their life. Cold-blooded animals have to eat much less than do warm-blooded ones. Sharks go for weeks, even a whole winter, without eating at all. In the Oceanarium, it is a real job to get them to eat a fish for the pleasure of the viewing tourists. They just keep cruising by; Liz has to stuff a fish in their jaws, which most of the time they refuse to bite. (And it is true that sharks do not like the taste of Homo sapien meat. It is surfers, stuffed into black wetsuits and lying on surfboards with feet in flippers, that, due to their poor eyesight, sharks sometimes mistake for seals, which some species of them do eat. They take a bite and then—like Count Dracula in Paul Morrisey's film deceived into thinking that this Italian girl in the 1960s he sank his fangs into is a virgin—puke it out when they realize their mistake.)

Each time a great shark careens by and pauses inches from my head, my eyes meet its small lemon-yellow eye fixed on me. Its ever-open jaws display rows of teeth. Sharks have skin like we do, not scales, but no expression. No tremors of curiosity, distrust, repugnance, antagonism, or menace shiver or crease that skin. I feel my eyes and my soul and big bloated body utterly exposed to that yellow eye, which reveals nothing whatever of its response to me. My eyes, unable to circumscribe, survey, foresee, cease to be eyes. Time ceases to pass from what had come to pass to possibilities, to a future, and extends in a motionless span, coming from nowhere, going nowhere.

14.7 CHTHONIAN EMOTIONS

Some 20 years ago I spent a winter in Kenya, visiting some of the Leakeys' excavations of bones of our ancestors from 4 million years ago. I wandered among the Masai, who have lived there for centuries, perhaps for millennia, without ever marking the earth. Herdsmen, they would not wound the earth to plant a garden or field, and live entirely off the milk, cheese, and blood drawn from their cattle. When the waterholes dry up, they do not dig wells but migrate to areas where there is water. They do not wound the earth to make mud bricks but live in huts made of cattle dung.

Then, after 4 million years of this, abruptly, just north, cliffs were dismantled and pyramids raised in Egypt. At once, in the first dynasties, the Pyramids were built, stoneworks on a titanic scale. This abrupt beginning of construction seemed to me a great mystery. The Pyramids appeared to be a great mystery to me. I could not imagine any society today building such monuments.

I returned to Cairo. I met a young refugee from the Sudan. He took me to where he stays. It was in the middle of a vast wasteland that had been the city garbage dump for centuries, so polluted that not a weed was to be seen anywhere. Here the most destitute of Cairo's poor make bricks and pots out of the muck. The terrain was pitted to hold ponds of water and adjacent ponds of wet clay, and the brick-makers and potters themselves lived in caves excavated in the muck or mud-wall huts covered with scavenged sheets of rusty metal. Their mud brick kilns were above and below these hovels, the black smoke seeping from them hovering low overhead. It was like ground-zero after the end of urban civilization. The substance of the muck was everywhere up against them, all their movements held in it, their glances and their touch ending in it, their thoughts inescapably terrestrial. I cast only oblique glances at them, apprehensive not only of the aggressions against foreigners that had been so much in the press but also of the resentment of the downtrodden everywhere for their homes and their destitution to be stared at by well-to-do outsiders. But we went inside one of these hovels. My friend's mother, to welcome me, poured from a jug some water in a cup of clay. I suddenly thought I had been transported to ground-zero where civilization began—for the first thing humans made with their hands were pots fashioned of clay. Cupped hands lifting water to the mouth were the first pots, and the first pots of clay were made to hold that gesture.

Beyond this wasteland, I could see the dunes of Giza and the Pyramids. The next day I went there. Everyone has seen them so often, in pictures in books, in films, in news broadcasts, in ads; the images and impressions collected on the surfaces of my eyes, ears, and skin while wandering among them had already all been projected there many times. Beyond these images and impressions, I tried to sense what the Pyramids are. In grade school, and in the books I now read, they were identified as tombs of kings who had made themselves divine—colossal monuments of a monstrous excrescence of egoism. That made them only the more unintelligible, and besides, it is surely false; one could just as well describe a medieval cathedral as a tomb for a lord bishop or king, on the argument that their mausoleums are found in them. In the beginning the pharaoh was not even buried in the Pyramid but in a tunnel in the bed rock outside of it.

The colossal constructions that are the Pyramids are some kind of cosmic markers. Their essential function is to do something in the cosmos. But virtually all the coordinates of the cosmos in which they are set are incomprehensible. The Nazca lines in which they are set have been effaced by the shifting dunes.

When the midday heat drove the tourists to their buses and the air-conditioned restaurants, I headed for the entrance of the Great Pyramid of Khufu. I was able to spend a few hours alone inside, until the tourists returned. The burial room is astonishingly small, bare, unsculptured; it is clearly incidental to the great functions of the monument (it is not even at the real center). In European cathedrals the burial chapels of kings and lord bishops are ostentatiously set out with statues of them along with heavenly angels and weeping women, and exposed to the veneration of the people. Here the now-empty sarcophagus is a plain stone box. For us now to identify the Pyramids as tombs of omnipotent kings is to view them as did the barbarian grave robbers. Not the last of these barbarian chieftains was Howard Carter, who sold half the plunder from Pharaoh Tutankhamun to the New York Metropolitan Museum for £14,000. In the burial chamber of Khufu all one sensed was the enormous reality of the stone about and above and below.

The Pyramids are nothing but stone, enormous stones in enormous numbers. Napoleon calculated that there was enough stone in the Great Pyramid of Khufu alone to build a wall three meters high around France. The Pyramids were not built by slaves, but by the whole population during the flooding of the Nile when the population was left unoccupied and when alone the river was high enough for the stones to be brought down from the quarries at the upper cataracts. The stones were not cut at the quarries in uniform blocks; they are of varying sizes and their sides of varying angles. They were cut to fit one another at the site, and they were fitted together so exactly that one cannot slip a wedge between them. They were fitted together that painstakingly not only on the outer face of the Pyramid, but in all the tiers, all the way through. This kind of absolute devotion to the stone really is inconceivable to us today. Something utterly transcendent, something of incalculable value, was sensed in stone, something to which all the energies, and years of life of a laborer could be devoted.

I went up the Nile to Luxor and the Valley of the Kings, where later pharaohs were buried deep underground. I came upon a tomb little visited because floods had pretty much destroyed the paintings, and once I entered I found myself alone. The tunnel is very long and steep, several hundred meters down. There is one landing halfway down, and then at the bottom a very large pillared room cut in the granite bedrock. In the center, the black sarcophagus is in mint condition. On the stone cover the face and folded hands of the pharaoh emerged, or were sinking into, the black stone. I am invaded by a sense of a human life coming to rest deep under millions of tons of rock. Here the surface agitations seem so remote from him, and, the few hours I linger there with him, from me.

I feel an imperative summons from the rock core of the planet to stay, and feel the serene immobility and majestic submission of the pharaoh an assignation.

Peace and fear—are these names for the emotions opening upon the depths of vegetative life, upon the skies, the ocean, the core of the Earth? Peace and fear are not the two basic emotional relations we have to nature; they are reflexive emotions, ways we feel ourselves, in our retreat or our lookout post, assured or threatened by nature. There is an inner momentum in emotion, to plunge ever outward. Toward the furthest biological and oceanic expanses, terrestrial depths, and celestial spaces of that geophysiological wanderer which is our planet.

NOTES

1. Bernard Williams, *Making Sense of Humanity* (Cambridge: Cambridge University Press, 1995), p. 238.

2. *Ibid.*, p. 239. "To me the human move to take responsibility for the living Earth is laughable—the rhetoric of the powerless. The planet takes care of us, not we of it. Our self-inflated moral imperative to guide a wayward Earth or heal our sick planet is evidence of our immense capacity for self-delusion. Rather, we need to protect us from ourselves....Our tenacious illusion of special dispensation belies our true status as upright mammalian weeds." Lynn Margulis, *Symbiotic Planet* (New York: Basic Books, 1998), pp. 115, 119.

3. Georges Bataille, *Erotism* (San Francisco: City Lights Books, 1990), pp. 21, 86.

4. Georges Bataille, *Visions of Excess*, transl. and ed. by Allen Stoekl: Univ. of Minnesota Press, Minneapolis, p. 13.

5. *Ibid.*, p. 13.

1 5

Four Ways to Look at earth

Peter Warshall

Peter Warshall is currently editor of the Whole Earth *magazine. His work focuses on conservation and conservation-based development. His background includes work in Ethiopia for the United Nations High Commission for Refugees, and with the Tohono O'odham and Apache people of Arizona. He has also served as a consultant to a number of corporations and municipal governments. His training includes advanced degrees in biological anthropology and in cultural anthropology, which help him to combine natural history, natural resource management, and environmental impact analysis toward the resolution of conflicts and the building of consensus between divergent interest groups.*

This essay approaches the earth sciences from a perspective that is typically neglected, that of the earth or soil. Warshall's goal here is twofold: to place the study of earth at the center of the Earth sciences, and to explore all the hidden dimensions of soil. Soil is expressive of a great range of human values: It is the source of fertility and medium of life, the material substrate for our engineering projects, a key player in various biogeochemical cycles, and the fount of our imaginative life. Soil is the living skin of the planet, the buffer and point of intersection among air, rock, and water, and the site of an incredible diversity of biota. Warshall calls for an earthy education that combines science and reverence for the substance we stand on and rely upon for our lives.

For a variety of historical reasons, earth itself ("soil") has never been central to the Earth sciences. Hans Jenny, the father of the soil sciences, once remarked that earth with a little "e" seemed to have become a taboo subject. The study of geological Earth had become divorced from the study of earth. Soil studies were for farmers. It was their medium to grow food and fiber, but soil was not part of or only a minor part of the geosciences. Others have speculated that earth, the foundation of real estate, frightened off scientists. They shied away from entangling themselves in the highly charged politics of private property rights and the cross-property management of erosion, water pollution, and wetland drainage. Still others consider that "earth is dirt," and dirt is no good. "Dirty hands," "dirty tricks," "soiled linen," "dirty minds," "to know the dirt about a friend" are all expressions that make the study of earth a bit repulsive to modern students.

Whatever the reasons, the study of soil within the geosciences has been fragmented, divided into overly specialized areas of expertise as well as alienated from ethical concerns that entangle soil with the future history of humans on the planet. Soil studies have been pigeonholed into geotechnical engineering, pedology, geohydrology, agricultural sciences, civil engineering (including sanitary engineering, highway engineering, dam construction), microbial biology, ecological restoration, meteorology, geobotany, geomorphology, and volcanology—to name just a few!

As soils are increasingly buried under concrete and asphalt, they become out-of-sight-out-of-mind. As kids in Brooklyn, we would watch the Water Department dig up the asphalt and speculate that, since this is where we saw the most dirt, it was where the city must store it. Very few citizens (or geoscientists) can answer what used to be simple questions: What soils did the food I am eating grow in? What is the name of the soil I am standing upon? What are its colors, textures, smells, even taste? Does my community have any special soils that cause problems (e.g., clays that swell and crack foundations of homes?) or enrich (e.g., special soils that support unique plants and animals)? Are my community's soils healthy? Where are the old industrial sites, landfills, or mine tailings that could be causing harm?

This essay attempts to jump-start a more holistic view of soil within the Earth sciences, emphasizing the multidimensional aspects of the living skin of our planet. I have approached soil in four ways:

1. *Soil as Malleable Material.* This is the engineer's view of soil; soil as a lifeless material to be shaped and moved to build our house pads, dams, and roads.
2. *Soil as the Medium of Life.* This is the ecologist's view. Soil is the living buffer of the planet, modifying air, water, geological and life cycles on all scales from farm field to troposphere.
3. *Soil as Fertility.* The farmer's view and the ancient view of soil as the source of our food, forage, and fibers. Soil as the Earth's flesh, parallel to water, which was, in older times, considered its blood.
4. *Soil as an Organizer of Society.* The source of various goods and services within human economic frameworks, especially as a driving force in human history.

The essay ends by pointing out the advantages of more explicitly incorporating ethics and a wider knowledge of soil into the biogeochemical sciences of the future.

15.1 SOIL AS MALLEABLE MATERIAL

The name Adam, the first human in Genesis, comes from the word for "clay," the malleable ingredient of soil that can be shaped into pots or bricks or sculpted images. All soils have shape-changing properties. They can be resistant or submissive to pressure, hard or loose, and sticky or flowing. Most builders and engineers view soil as the lifeless material near or at the surface of the landscape that requires reshaping to become useful to human endeavors. In this view, geologists sometimes call soils "unconsolidated materials" to distinguish them from the more massive and difficult to shape bedrock.

The central focus of the engineering view is, how much muscle does it takes to move or deform soils? Earth movers commonly discuss soils by the tools and machines required to reshape them: the shovel, plow, backhoe, compactor, ripper, scraper, or cat (bulldozer). Equipment operators become intimate with the strengths and weaknesses of various kinds of earthen materials. They develop remarkable skills in shaping them into safe, long-lasting foundations and structures on which modern society depends.

Home foundations, retaining walls, highway and railroad sub-bases, tunnels, dams, cellars, levees, and mines—the very infrastructure of modern civilization—derive from knowledge of the physical and mechanical personalities of various earthen materials. By friend and foe, the earth movers—reshapers of the planet's surface—are often viewed as playing God.

Viewing earth as a malleable substance demands a sensual attachment; a pleasure in squeezing, rubbing, rolling, wetting, poking, even tasting soils. "Experts" can feel out soil's cohesiveness, hardness, and adhesiveness—qualities necessary to work with soils in efficient, if not elegant, ways. *Cohesiveness*, for instance, defines soil's resistance to being torn apart. It depends on friction between the soil's mineral particles and plays a major role in keeping soils from sliding down hillslopes. When cohesive, a watershed's soils support a wealth of plantlife. For redwoods to survive thousands of years, the soil mass must remain in place. Clearcuts or road cuts can undermine the friction between soil particles and the hillslope. When this occurs, soils slide off the slope.

As opposed to cohesive soils, naturally loose soils are a major headache to geotechnical engineers. Much time and money has been spent to stabilize quicksands, sand dunes and muds, soils subject to frost heaves, soils that creep, clays that crack, or soils subject to earthquake liquefaction.

Hardness is the second quality of malleable soils. Hardness is not just a gardener's problem of forcing the shovel, toiling to penetrate a dry, clay loam earth. Hardness and penetration also pose problems for tree roots seeking nutrients and water, prairie dogs constructing subterranean towns, and worms seeking food. Farmers, though, pay hardness the most attention. They must plow, rip, and hoe hard soils to increase root and water penetration. In contrast, road and adobe home builders want hardness, especially a soil that will resist the hammering of rain drops.

The third quality is *adhesiveness*. Elephants and Californians taking mud baths, and beauticians giving facials, cherish adhesive soils. Sticky soil helps keep insects from biting and may draw toxins from the human skin. But adhesion is a problem for equipment operators. Wet mud gloms onto bulldozer blades and backhoe buckets, burdening work.

All these mechanical and physical aspects of soil ignore the life and microstructure within soils. Over the past quarter century, the earth movers have bumped heads with those who see value in the biological richness (see below) of the soil mass. Rather than just an "unconsolidated material," the newer view technically calls soil a "multiphasic material." Multiphasic soil has four components (water, air, solid particles, organic materials) and a mini-architecture of pore spaces that will change soil's strength under different circumstances. Of the four components, mineral grains and clays have been most familiar to geoscientists and earth movers. Unfortunately, the value of soil organic matter, the genetic resources (see next section), and the balance between air and water circulation within the soil's micro-architecture (sometimes called "ecostructure") have been largely ignored.

The multiphasic approach includes seeing soil-water and soil-air as life-supporting networks, not just as problems to building earthworks. An agricultural soil, for instance, might contain 50 percent empty pore space (to accommodate flows of water, nutrient, and air); about 9 percent organic materials (humus and roots), and about 40 percent solid mineral particles all intricately latticed. After heavy farm equipment passes over the field, the pore spaces can compact. The topsoil begins to erode because the architecture of the layers has been condensed. The rain—instead of seeping into the soil—runs over the surface. In addition, the organic matter decreases because roots can no longer penetrate the compressed pore spaces. While this is the "same" soil from the point of view of "unconsolidated materials," it is a very different soil from the point of view of fertility, erosion, and sustainable agriculture.

The goal of earth guardians is to keep a soil in place and to retain its organic component and pore spaces available to air, water, and growing organisms. The conflict between earth movers and earth guardians can become intense. For instance, during the California gold rush, hydraulic miners unearthed gold by blasting apart riverbanks with high-pressure hoses. The soil became the silt and sediment of the watershed's channels and ruined many of California's once great salmon fisheries, as well as leaving the hillslopes bare of topsoil. Now, the geosciences (formerly concerned with the mineral ores or the gravel and sand required for construction) have begun to consider themselves part of conservation biology, the health of fisheries, as well as watershed stability. Being specialized is no longer logical nor tenable.

The organic component especially has gained importance, even in the eyes of the building industry. In many housing subdivisions and mine developments, for instance, operators stockpile the topsoil, which is the most fertile component of soil and the most able to grow plantlife. After reshaping the harder subsoil to better accommodate building foundations or mining, operators replace the topsoil in appropriate locations. They have tried to accommodate the biologist's perspective and understand that various layers of the soil profile have different life-supporting values. Soil is more than just tight or loose.

This worldview change in the geosciences is not just a change from an isolated science to an interdisciplinary science. The ethical values of geoscientists, geotechnical engineers, construction contractors, and the land-use planners evolve as they incorporate the life-supporting aspects of soil into their worldview. From a historian's and anthropologist's perspective, the old characteristics of soil—cohesion, hardness, and adhesion—were also metaphors of what was valued in civilization: a strong, cohesive society that adhered to prescribed rules. These attitudes were reenforced by human activities such as constructing roads, cultivating lands, building pyramids and cities and convincing citizens that these big projects were good for society as a whole. Perhaps it is not a coincidence that the priests, politicians, generals, and governors—those who ponder how to "shape" society—adopted the same vocabulary as that of the builders of its infrastructure, a vocabulary of the mechanical qualities of soil. It is hard to "prove" this historical assertion, and historians have yet to research documents that might support or refute it. Conversely, with the recent concern for a sustainable Earth, the geosciences have already begun to add "life" (organics) and life support (air and water) to their perceptions of soil. This change will integrate biology with geology and perhaps help shift the meaning of governance in postindustrial civilizations.

15.2 SOIL AS GAEAN BUFFER

In a dramatic contrast to geo-engineers, ecologists primarily understand soil as the living skin of Earth. The "pedosphere," the component of the entire planet that contains its soils, nurtures the greatest number of biological organisms, the most numerous biochemical transformations and kinds of molecules, and the most diverse biophysical transformations on any volume of the planet. Moist soils display more rapid-fire exchange of electrons with more pathways and alternative connections than the human brain or the highest-powered parallel computer. It is the home of the most intimate associations (especially as colloids) between the organic and inorganic, and biotic and nonbiotic.

> One teaspoon of good grassland soil may contain 5 billion bacteria, 20 million fungi, and 1 million protoctists. Expand the census to a square meter and you will find, besides unthinkable numbers of the creatures already mentioned, perhaps 1,000 each of ants, spiders, wood lice, beetles and their larvae, and fly larvae; 2,000 each of earthworms and large myriapods (millipedes and centipedes); 8,000 slugs and snails; 20,000 pot worms, 40,000 springtails, 120,000 mites, and 12 million nematodes. (Wilson 1984)

This view is the opposite of soil as unconsolidated material. It focuses on soil as the living skin connecting the microcosmos of one teaspoon of soil to the macrocosmos of the planet's largest-scale cycles of water, heat, greenhouse gases, nutrients, sediment, and breathable air. Soils link the micro- and the macro-realms by circulating and biochemically transforming the planet's nutrients and water; by collecting nutrients, water, and wind-blown dusts into the soil body; by storing and retaining water, humus, and toxic chemicals; by filtering nutrient and toxins through various soil "treatment facilities," and by dispersing and scheduling the flow of water, nutrients, and toxins into the air, through the soil mass or into streams and rivers.

Microbial soil bio-cycling, for instance, transforms all dead organisms back into usable nutrients or stores them in long-term "safe deposit boxes" called *humus*. The humus ties up carbon as complex molecules and keeps it from returning into the atmosphere as one of the major greenhouse gases. The more humus, the lower the risks of global warming. Plowing the soil can break open the carbonaceous safe deposit boxes, and release a significant amount of greenhouse gases back into the atmosphere. In this view, stewardship for the life qualities of soil, teaspoonful by teaspoonful, significantly contributes to planetary stability.

Similarly, lathering the earth with asphalt stops the cooling abilities of soils and their plantlife, and impedes rainfall from infiltrating into the soil. Atmospheric temperatures increase (especially around cities, which can be viewed as "heat islands"), and subsurface flows within soils and fractured bedrock to rivers and reservoirs decrease. Pollutants wash quickly off the city streets into water bodies, and signs that prohibit swimming pop up around beaches and lakes. Urban hard surfaces prevent rain from entering the soil body where microorganisms can bio-cycle and transform pollutants into harmless substances. Again, building infrastructure is more than plastering over unconsolidated materials. It shuts out the free services of cooling, water storage and water treatment provided by a living earth.

Ever since James Lovelock and Lynn Margulis presented the Gaea Hypothesis, a major change has occurred in the Earth sciences. The Gaea Hypothesis states that life has had a major impact on everything from climate to plate tectonics, and that life acts as a buffer to keep the planet from fluctuating to the extremes found on nearby planets such as Mars and Venus. Soil is rapidly emerging as the central organizer of all the planet's largest-scale components: the atmosphere, the hydrosphere, the geosphere (or

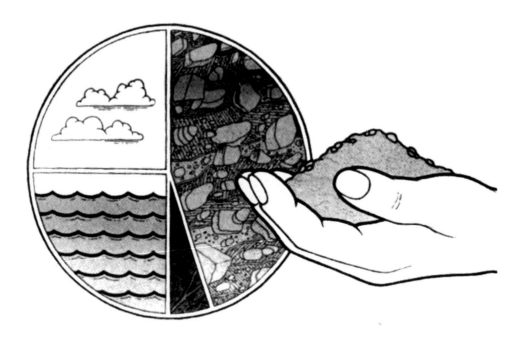

Figure 15.1 "Good" soil is considered 50% empty pore space (to accomodate the flows of water, air, and nutrients); about 9% soft-textured organic materials (humus); and about 40% solid grain particles. The soil is intricately latticed.

lithosphere), and biosphere. The pedosphere, or soil component (especially if the upper sediment layers of freshwater and ocean bottoms are included), appears as the Earth's central ecological node where all the "spheres" link up (see Fig. 15.1).

Geoscientists feel uncomfortable with calling their science after the Greek Goddess Gaea. Instead, they call essentially the same scientific pursuit, Earth Systems Science. The Gaean perspective has therefore been unsettling to classical geoscientists. Almost all near-surface "geochemical" processes have been found to have a living organism involved. Mobilization, sequestering, regenerating, scheduling, and channeling of minerals and inorganic molecules cannot be untangled from life's chemistry. Not just coal and oil, but also ores like iron are dependent on the presence of life (banded iron deposits required bacteria-produced oxygen in the early atmosphere). In other words, geoscientists are more accurately biogeochemical scientists. And the geosciences are more accurately biogeochemical sciences: *bio* means "living," *geo* means "earth," and *chem* means "from-the-elements."

For instance, recent U.S. Geological Survey publications (1993, 1998) talk of understanding the geochemistry of our fragile environment and welcome the new field of environmental geochemistry. However, another step is needed that more explicitly incorporates *bios*. For example, one technique used to restore toxic mine tailings is "phytoremediation." Specific plants can absorb heavy metals. Mining companies plant them on the mine tailings where they concentrate the toxins in their stems and leaves. Mine workers harvest and burn the plants to recover the metals or place them in safe landfill sites. Phytoremediation can be a lot cheaper than mechano-chemical techniques to clean up mine wastes, but for the living plants to be effective, we must have a deep understanding of the living soils. At the complex interface of soil particles—covered in microbial "slimes," entangled among plant rootlets, and laced with fungi that both penetrate the plant roots and surround the soil particles—phytoremediation begins. Living creatures star as the crucial processors of chemical transformations.

To give another example, special bacterium called "ultra-bacteria" can be injected into certain soils and fractured bedrock where they multiply and expand their body size. The bacterial colonies grow into a bio-barrier that slows or prevents the movement of polluted groundwater. This is pure biogeology.

In this new view of soil, it is not only the spatial connection of the micro- to macro-realms that is important. Soil plays a central role in the speeds at which both local watersheds and the planet can heal themselves. At the watershed scale, deep, absorbent soils can retard the flow of water and prevent severe flooding, or store toxic pesticides long enough for microbial breakdown. On the planetary scale, reduced plowing or untouched forests may be methods to slow the buildup of greenhouse gases in the atmosphere. (The soil will store more carbon-based matter if left undisturbed.) In both examples, the "pedosphere" acts as a bio-valve slowing or speeding the flows of nutrients, toxins, gases, and all other essentials needed for existence. Along with water, it is the only landscape component that can be easily managed for the benefit of humans and the planet.

Compare the speed of rate changes with the major components of the planet. Bedrock forms the slowest-moving component of the Gaean system. Water, even in the form of ice, is less solid and more viscous. Both rock and water change from solid to liquid to vapors. Rocks, of course, change from solid bedrock to liquid magma or volcanic flows or sulfurous fumes on longer time scales. But both the geosphere and the hydrosphere turn over at much slower rates than does air. The composition of air changes quickly and drastically from, for instance, its claustrophobic stale air in pockets of soil to its dispersal by the winds and upper layers of the atmosphere. The pedosphere mediates the speeds of the planet's rock, air, and water cycles somewhat like the clutch in a car.

The micro- to macro-understanding of biogeochemical cycles, and how their rates of change differ by season, location on the planet, and interlocking feedbacks will require immense scientific study, which is of pressing importance. For the first time in recorded history, humans have surpassed the Earth in contributing to the flow of materials. More soil is eroded, more nitrogen moved, more heavy metals and carbon released, and more heat generated by humans than by natural planetary processes. The old-style geosciences undervalued living soils and the impacts of human beings on the cycles of the whole planet. The geosciences are not what they might have hoped—the all-encompassing planetary science. But, with the addition of life sciences, the "biogeo" remains a significant piece in discovering and managing the planetary life-support puzzle.

15.3 SOIL AS FERTILITY

One particular quality of soil stands out for humans: its fertility. Ever since the Neolithic, humans have felt so close to soil that they gave thanks, offered prayers, and performed ceremonies to encourage soil to be fertile, to produce abundantly, and to persist in its generous life-support. In China, earth was considered the mother of all things. It was the flesh, just as water was the blood. Our flesh and plumbing were part of the planet's flesh and blood. In ancient Japan, the emperor had to make love to a virgin on the open fields before the fields could be planted. In Europe, a night of free love occurred to mimic and to celebrate the thawing of the winter soils. Adultery was not permitted, except on these special nights, when, to honor the now fertile earth, it was decreed. The Old Testament spends many verses not, as commonly reported, declaring human dominion over the soil, nor encouraging land stewardship, but more radically stating that humans should be servants of the soil.

All these ancient ceremonies combined soil and morality. The good life comes from devotion to and respect for the soil. Respecting the limits of a soil's fertility teaches humility. Cultivating and maintaining soil fertility teaches humans responsibility and moral flexibility. Accepting larger forces like floods and droughts that undermine soil fertility nurtures humility. These values have been lost by humans in postmodern industrial societies, whose daily lives no longer involve working with soil. All these words—*humble, humility*, even *humane*—have become fossilized. Only their etymology, their origins, unearths their buried meanings. "Humane," "humble," and "humility" all come from the same Indo-European word root that means "humus," or fertile earth. We cannot overestimate the linguistic alienation of geoscientists as well as the public from this kind of moral attention to the fertility of soil.

Today, financial values have largely replaced earth-derived values. From the point of view of economics, fertility is a measure of financial value. The story told is: The greater the fertility, the greater the yield, the greater the income. We are not servants of soils nor their students. Soils serve as inanimate materials made to yield more and more by manipulation and scientific pursuit, not by ritual and intuitive knowledge. Humble and awesome feelings about the floods that brought fertility by spreading silt have been challenged by levees for flood control and petrochemical fertilizers that replace soil organic matter and nutrients. Indeed, in some orange orchards on sandy soils in Florida, soil is simply a prop to hold the trees upright. The oranges are fruit "manufactured" by irrigation water and petrochemical nutrients supplied by industrial agriculture. Stop the energy-intensive flows of pumped water and fertilizers and the soil returns to near-sterile sand.

To maintain a fertile soil and pass it on to the next generation, the modern farmer has no need to copulate on the fields in early spring. But, farmers and geoscientists can return to seeing value in perpetual soil fertility by contemplating its five components: the amount and quality of soil organic matter (SOM), the condition of the soil's ecostructure, the soil's moisture-holding capacity, its mineral grains and clays, and its genetic resources. These components can be considered the soil's Fertility Portfolio, a portfolio that describes the true capital value of the soil. The Florida sandy soil has no fertility value without inputs. A truly rich soil will have a portfolio with high SOM: an intact, open ecostructure, a water-holding capacity that is resilient to drought, a balanced mixture of soil particles and clays, and great biodiversity. Here is the briefest summary of the Fertility Portfolio.

Soil organic matter is the blue chip stock of fertility. More humus means more bio-available nutrients for plants and microbes; SOM buffers the soil mass from outside threats to fertility and productivity. With greater SOM, the soil performs better under stresses like drought or floods.

Soil ecostructure is the three-dimensional lattice built by aggregates of soil particles and bioslimes and by the layering and mixes of clays, sands, silts, and humus. It is the architecture or the physical structure that keeps pore spaces from collapsing and serves as the conduits for nutrients. The ecostructure also supplies the interfaces where nutrient exchanges occur. These ecostructure sites can be considered like micro-ATMs or bank tellers that provide friendly efficiency to microbial and vegetative customers. Nutrient flows (the "cash currency") decrease when the soil cannot move funds (e.g., a compacted soil with fewer exchange). Or, the nutrient currency may move too fast out of the soil's savings account if, for instance, erosion or plowing destroys the ecostructure. Cherishing the fine intricacy of soil ecostructure and fertility explains, in part, why Native Americans became horrified at strip mines, and told miners they were tearing apart their mother's breasts.

Soil water provides the "liquidity" for nutrient fund movements. Soil moisture also creates the right environment for microbes to restructure "fixed assets" like cellulose and lignin. Soil water leaches out salts, which, if left in the bank, would corrode humus formation. The main instrument for keeping the portfolio balanced is internal soil water.

Minerals, clays, and weathered rock provide the anchorage for plants. Their sheer weight helps hold the soil fertility complex in place. Mineral particles also provide many of the surfaces for nutrient exchanges.

The genes of soil microbes and larger organisms like worms and moles issue the instructions to convert all kinds of fungible nutrient instruments from one to another. Each creature's waste products become the food of another creature. The result is the maintenance of in-place soil fertility.

The genetic structures of crops play a most important role in setting the rules for nutrient withdrawals. Plant-by-plant the genes define the required nutrient needs. Plant-by-plant the genes define what artificial inputs (fertilizers and imported water) must be added to the soil. In the coming century, gene-splicing of microbes, crop plants, and insects will alter every aspect of how we think of soil fertility. We could dream, for instance, of a worm that eats more mineral particles faster and, by doing so, restores ecostructure to a compacted soil. Or dream of nitrogen-fixing genes attached to corn that will reduce the need for petrochemical fertilizers. Or possibly researchers will construct a microbe that breaks down toxic herbicides that remain within a soil.

In summary, the artificial fertility of human-manipulated soil requires imports of additives. In addition to crop genetics, mining mineral fertilizers such as phosphorus and potash, and transforming ever-increasing amounts of petroleum into urea-based fertilizers, has been the basis for increased yields in the second half of the twentieth-century. Ultimately, these "ores" will become the limiting natural resources of artificial soil fertility. In the twenty-first century, more and more petroleum will be directed away from transportation and industrial fuels and toward petroleum-based agrochemical products required to feed the human population. But the Fertility Portfolio becomes unreliable when it accepts new minerals or chemicals that distort the natural processes that create soil fertility. Pesticides or heavy metals may kill both the microbial bank tellers and the targeted pest. Artificial fertilizers alter the soil's structure by changing the chemistry of the soil and by "addicting" the soil to these additives.

The new sense of "sustainable" soil fertility places new demands on Earth scientists. Soil, once banished to agricultural schools, must now take center stage in the geosciences. Old earth words and values, within modern scientific contexts, will regain new life.

15.4 SOIL AS AN ECONOMIC GOOD AND SERVICE

Soil is a relatively uncelebrated driving force in human history, especially economic and political history. Without fertile soils, humans must trade or go to war for goods grown exclusively from fertile soils. I like to tell the soil-centered version of Athenian history, starting from the invention of the metal ax; followed by the cutting of the sacred oaks, and then the erosion of the soils of Hellas. With topsoil washed away, the Greeks could no longer grow sufficient grains. They had to build a navy to conquer and trade with societies that had retained their topsoil and their grain crops. They had to learn to use their subsoils, which had become the "top" soil. These more clayey layers could not grow grains but could grow olive trees and grapes. They mined the clayey soils for amphorae (vases to store and transport olive oil and wine). The decoration of the amphorae accelerated the blossoming of Greek pictorial art. The whole of Hellenic civilization was transformed because it lost its topsoil. Athens fought for 28 years with Sparta, which had retained its topsoil and guarded its granaries. The Athenian period eventually ended as its navy and trading vessels became vulnerable to other nations' warships and it could no longer thrive on other people's topsoil crops. Contrast this telling of history—with soil as the central character—with the stories that celebrate Pericles and other heroes as the major driving forces of Athens' rise and fall.

It is easy to forget that soil "bankruptcies" ended many civilizations. Loss of fertile soils from erosion or salt catapulted the civilizations of the Tigris-Euphrates, the Yangtze, today's Amazon, and others into most difficult times. The Dust Bowl of the 1930s threatened to close down America's Midwestern heartland. It later was at least temporarily rejuvenated by a change in weather and by the propping up of its soils with petrochemical fertilizers and pesticides. But its history like so many others is a tale of soil use and abuse.

Soil, then, is the crucial part of both local and global economic systems. It provides both critical human goods and human services. Soil-derived goods include pigments for paints, cosmetics, penicillin and streptomycin, antidiarrhea and anti-indigestion medicines, gravels for landscaping and construction, sand for cement and glass, artificial turf, adobe and brick for homes, and clays for pottery and high-quality filters.

Many folks, for instance, ate and still eat certain soils. Geophagy (earth-eating) may, in fact, be millions of years old, as a specific white clay has been found with the bones of *Homo habilis*, a clay not native to the area in which the bones were found.

Today, we drink kaolin clays in various antidiarrheal medicines. Many pregnant women all over the world ingest particular clays to improve the nutrition of their fetus. The Ainu people of Japan have a special recipe for clay soup. It is a good-tasting hunger-reliever. Others eat clays to buffer poisons. In Peru, for instance, certain wild potatoes contain toxins. By dipping them in a bowl of liquid clay, the toxins are neutralized. (Peru is the source of potatoes and contains hundreds of wild varieties in the very poisonous Nightshade Family.) From Finland to the United States, grandparents can still tell stories of "clay breads" used to both allay hunger and gain mineral nutrients. Because both streptomycin and penicillin come from soil, there is the possibility that eating certain soils were known to heal, an intuitive understanding that they contained antibiotics. Few Western scientists have studied geophagy, perhaps because it seems ridiculous to us to eat dirt.

Soils provide services such as purifying water in watersheds, humidifying the air, anchoring living things, buffering the atmosphere, storing carbon, acting as the central nutrient exchange for food webs, and marketing electrons between living beings and mineral particles. The economic value of these services has been conservatively calculated in the trillions of dollars. The recognition of these services has caused a shift in emphasis within the geosciences from the goods of earth to its beneficial processes.

Soil remains, for instance, the essential "free" fertility service in food, fiber, and forage production. Hydroponics, the only nonsoil-based food system, is very expensive and not yet reliable for large-scale production. Although many petrochemical fibers like nylon and kevlar have replaced earth-grown fibers like cotton, humanity's dependence on natural fiber from soils has not diminished. All our meat comes directly from grazing on poorer soils or production of feed grains on fertile soils.

Any good or service that participates in the human organization of trade is bound to be political: Power interests struggle to make the most profit at each stage of the system as governing bodies try to enforce a more equitable distribution of costs and benefits. Soil is no exception to politics, as it serves to organize our food system. Take the richest soils in the world, the soils of the upper Mississippi River basin. These soils connect the corn, soybean, and wheat farmers to the grain elevator operators and buyers to the processors, the wholesalers and retailers and your dinner plate. To maximize financial profits (rather than soil wealth), many farmers have tried to displace environmental costs onto other citizens. They have been forced or tried to escape responsibility for long-term soil wealth. In the upper Mississippi, for instance, one-third of the soil organic matter has been lost from its farms. Soil erosion has silted in the river channels, and taxpayers pay for dredging to help subsidize the barge industry trade in grains for export. Petrochemicals have been washed from the farms downstream, forcing cities to spend their tax money cleaning up the river water for safe drinking. The lower Mississippi is known as "Cancer Alley." By the time the silt, excess fertilizers, and toxic pesticides reach the mouth of the Mississippi, they settle into a 7,500 square mile area called the Dead Zone because no commercial fish or shellfish can survive in this damaged environment. In other words, a political battle ensues. Who should pay for damages to the river channel, drinking water, and lost income to fisherman? Does not responsibility and externalized costs start with an attitude toward soil fertility and crop yields?

The government has tried to intervene to reduce costs to downstream peoples, keep the peace, and, among a few politicians, take responsibility for soil fertility for future generations of farmers. The taxpayer, through Congress, pays for wetland reconstruction and protection to compensate for river channel and on-farm damages. The taxpayer also pays for income supports for farmers willing to remove their land from production so as to reduce soil erosion. But downstream citizenries have not been satisfied by government interventions on their behalf. They have begun to sue upstream citizens to force them to take responsibility for the water quality leaving their land. In contrast, the farm lobby has pressured Congress to prevent stricter standards that would force farmers to reduce excess fertilizer and pesticide discharges from their lands. The questions remain: How much outside regulation should there be over farming? How much are farmers responsible for damages that occurs off their farms? This is the heart of contemporary soil politics in the United States.

The geosciences can contribute to resolving these conflicts and building a consensus by helping farmers work with their soils, maintain yields, and reduce silt and water pollution. But the skills required will be more than water chemistry or pedology (the study of soils). Conflict resolution moves geoscientists from "neutral" providers of information into society at large. Most geoscientists have not been trained in consensus building, balancing ethical values, risk analysis, and reaching compromises. However, once soil is recognized as a crucial economic service, geoscientists cannot but plunge into the political world.

Soil provides many other services besides food production. These include buffering the waters of the planet from acid rain; storing carbon to reduce greenhouse gases and global warming; filtering and treating irrigation and municipal wastewaters as well as recharging and purifying groundwaters for drinking water. A land and water ethic that connects the geoscientist's knowledge with a sense of desired conditions for future generations may well return us to a consideration of "ancient" values.

15.5 CONCLUSION: SOIL AND KNOWLEDGE

Our present educational system does not give much value to soils as a subject. Rarely do either biology or geology classes give them significant attention. Most college texts on the Earth sciences barely mention them as biological phenomena. There is no holistic text on soil. We have seen how fragmented the knowledge of soil has become and how it has been boxed off into many professions.

Recently, educators and soil practitioners have made small steps toward recovering a more comprehensive view of the ground beneath our feet. New visions that jump boundaries are, for instance, microbial ecology of soils, ethnopedology, and applied geomorphology.

Microbial ecology is the foundation—the natural history. As E.O. Wilson, one of America's greatest biologists, has observed:

> If I could do it all over again, and relive my vision in the twenty-first century, I would be a microbial ecologist. Ten billion bacteria live in a gram of soil, a mere

Figure 15.2   There is an excellent interface between soil and roots in this corn plant. For the crop speculator, the nutrient (currency) flows will efficiently generate natural capital (root mass, organic matter, and corn). She will convert this knowledge of natural capital formation into artificial or financial capital (e.g., corn futures at the Chicago Board of Trade).

pinch held between thumb and forefinger. They represent thousands of species, almost none of which are known to science. Into that world I would go with the aid of microscopy and molecular analysis. I would cut my way through clonal forests sprawled across grains of sand, travel in an imagined submarine through drops of water proportionately the size of lakes, and track predators and prey in order to discover new life ways and alien food webs. All this, and I need venture no further than ten paces outside my laboratory building. The jaguars, ants, and orchids would still occupy distant forests in all their splendor, but now they would be joined by an even more complex world virtually without end. For one more turn around I would keep alive the little boy of Paradise Beach. (Wilson 1984)

"Ethnopedology" is a recent combination of anthropology and soil science in which investigators try to learn how societies classify and value soils. Western cultural soil scientists, for instance, have built a taxonomy based on the size of mineral grains. Sand is the biggest, then silt, then clay. They classify soils by the proportions of sand, silt, and clay. They then overlay this taxonomy with the clumpiness of the clods. These "aggregates" can be blocky, platy, or prismatic. They then add other characteristics such as acidity or saltiness, which further categorize each soil. Other cultures—for instance, French scientists and Zunis—have built their taxonomy more on their perceived history of the soil. Did it come from wind, water deposition, in-place weathering, or biological decay? In Haiti, farmers distinguish between "hot" and "fat" soils. Soils that are "hot" tend to dry out and "desertify" quickly. Fat soils are very fertile and can make you fat from eating the food produced. Some cultures make more distinctions than do modern soil scientists, and some make fewer.

Ethnopedology combines cognitive science, epistemology, and the earth. A well-intended soil scientist, visiting an African village, may give poor advice because he or she does not understand how the local people classify and think about soils. In Ethiopia, for instance, many peasants have two farms—one with soils that produce in good rains, and one that produces something no matter how little rain. Development "experts" thought it was silly to have the peasants walk so far between fields and tried to encourage farm consolidation. Little did they understand that, in dry years, this put the peasants in greater danger of no crops and family starvation. A clear understanding of soil names would have brought the Ethiopian agro-soil knowledge into harmony with the good intentions of the experts.

In the American Southwest, many ranchers and conservationists wish to restore grasslands, but experiments with grassland restoration have not been very successful. Recent study has begun to combine geological history with soil formation and landscapes. This field of applied geomorphology (literally, "earth-shape") has proved immensely helpful to restoration ecologists. By dividing the landscape into landforms, geomorphologists have mapped why one area has clayey soils and another carbonate (chalky) soils. They have mapped which soils lost their topsoil and now have a "subsoil" as the top horizon or which soils have been buried under landslides. They work with botanists and range experts to determine which grasses and forbs do best on which landscape surface. This helps prevent fruitless labor such as trying to reseed an alluvial fan with seeds that will not grow there or trying to grow topsoil grasses on a soil whose topsoil has been buried or eroded away. The studies also nurture humbleness as geomorphologists and land users come to appreciate that it took from tens to hundreds of thousands of years for nature to shape and create the landscape they now live upon.

But the geosciences and related disciplines are still a long way from a reverence for soil. No geological society, for instance, has proposed an Endangered Soils Act as a technique to preserve remnants of this continent's original soils. There are over 9,500 recognized soil types in just the United States. These remnants of virgin soils would serve as the "baseline" for comparisons of degradation or improvement of the nation's soils. The outdoor "museum" soils could educate the public about the difficulty and expense of replacing what nature has made. For soil lovers, they would become stops on a pilgrimage or tour of nature's shrines.

The now largely urban public needs to renew its interest in soils as a precondition for a new moral groundedness. An education that starts with a reverence for the substance we stand on and rely on for our life can help shift values from financial wealth to soil fertility, and, perhaps, even rekindle interest in humility and humane (self-limiting) manners and behavior. The task is enormous. The urban public has little opportunity to dig soils, squeeze them, and think about the meaning of their layers, colors, smells, and taste. This outdoor involvement with soil was a primary educational experience for humans for millions of years. Sensual contact with soil, especially walking barefoot, was historically one of our species most informative lifetime experiences. Perhaps with a revival in outdoor education, intellectual understandings, philosophical musings, and morality we can still germinate a soil-sustaining ethics from an earthy base.

REFERENCES

Alexander, Martin. 1977. *Introduction to Soil Microbiology*. Melbourne, FL: Kreiger.

Brady, Nyle, and Raymond Weil. 1999. *The Nature and Property of Soils*. Upper Saddle River, NJ: Prentice Hall.

Dailey, Gretchen. 1997. *Nature's Services: Societal Dependence on Natural Ecosystems*. Washington, D.C.: Island Press.

Groning, Karl. 1998. *Body Decoration: A World Survey of Body Art*. New York: The Vendor Press/St. Martin's Press.

Hillel, David. 1991. *Out of the Earth: Civilization and the Life of the Soil*. Berkeley, CA: University of California Press.

Houben, Hugo, and Hubert Guilland. 1994. *Earth Construction: A Comprehensive Guide*. Herndon, VA: Stylus Publishing.

Logan, William Bryant. 1996. *Dirt: The Ecstatic Skin of Earth*. NJ: Riverhead Books.

Margulis, Lynn, and Dorion Sagan. 1986. *Microcosmos*. New York: Simon and Schuster.

Nichols, Herbert. *Moving the Earth: The Workbook of Excavation*. New York: McGraw Hill, 1998.

Oldroyd, David. 1996. *Thinking About Earth: A History of Ideas in Geology*. Cambridge, MA: Harvard University Press.

Oregon State University. *Manual for Judging Oregon Soil*. Corvallis, OR.

U.S. Geological Survey. 1993. *Understanding Our Fragile Environment: Lessons From Geochemical Studies*. Circular 1105.

U.S. Geological Survey. 1998. *Geology for a Changing World: A Science Strategy for the Geologic Division of the USGS, 2000–2010*. Circular 1172.

Whole Earth. 1999. Special issue: "Celebrating Soil—The Mother of All Things," no. 96, pp. 4–45.

Wilson, E.O. 1984. *Naturalist*. Washington, DC: Island Press/Shearwater Books, 1994.

Index

This index to persons bypasses the preface and the list of references sometimes found at the end of individual essays, but includes notes, which are indicated with the letter n. The author of an essay is indicated by ff following the opening page number, which is normally the only page on which the author's name actually appears.